# Lecture Notes in Business Information Processing     **462**

Ērika Nazaruka · Kurt Sandkuhl ·
Ulf Seigerroth (Eds.)

# Perspectives in Business Informatics Research

21st International Conference
on Business Informatics Research, BIR 2022
Rostock, Germany, September 21–23, 2022
Proceedings

Springer

*Editors*
Ērika Nazaruka ⓘ
Riga Technical University
Riga, Latvia

Kurt Sandkuhl ⓘ
Rostock University
Rostock, Germany

Ulf Seigerroth ⓘ
Jönköping University
Jönköping, Sweden

ISSN 1865-1348                    ISSN 1865-1356 (electronic)
Lecture Notes in Business Information Processing
ISBN 978-3-031-16946-5          ISBN 978-3-031-16947-2 (eBook)
https://doi.org/10.1007/978-3-031-16947-2

This Springer imprint is published by the registered company Springer Nature Switzerland AG
The registered company address is: Gewerbestrasse 11, 6330 Cham, Switzerland

# Preface

Business informatics is an academic discipline that addresses the development, combination, and integration of concepts, approaches, and technologies from computer science, information systems, economics, and business administration with a focus on applications in enterprises, public authorities, and other organizations. Business informatics is closely related to the area of business information systems.

The conference series on Business Informatics Research (BIR) was established in the year 2000 as the result of a collaboration of several Swedish and German universities. The goal was to create a forum for the discussion of latest research results, new research directions, and PhD topics in the business informatics community. The conference has expanded to many other countries, and usually includes workshops as well as a doctoral consortium accompanying the main conference track. The BIR Steering Committee currently comprises 17 members from 13 countries, many of them organizers of previous BIR events. The 21st International Conference on Perspectives in Business Informatics Research (BIR 2022) took place in Rostock, Germany, during September 21–23, 2022, at the Institute of Computer Science of the University of Rostock. The local organizer was the institute's Business Information Systems research group.

The central theme of BIR 2022 was "Business Informatics for Sustainable Innovation". Digital transformation, artificial intelligence, the Internet of Things, and dataspaces are some of the current trends in business and market environments that lead to innovations in products, services, organizational structures, and work environments. These innovations should create value for all stakeholders involved, not only from a business and commercial perspective, and not purely based on technology, but also from a social or environmental perspective. The topic of sustainable innovation also addresses innovations tackling the global climate emergency, transition to a net-zero economy, or changes in ecosystems. Relevance and importance of business informatics for sustainable innovations has been recognized by an increasing number of companies and public agencies. Achieving sustainability requires a multi-perspective approach taking organizational, economic, and technical aspects into account. In a world of cloud, social, and big data, additional challenges for business informatics and the design of information systems architectures are introduced, e.g., in respect of the design of data-driven processes or processes enabling cross-enterprise collaboration. To deal with these challenges, close cooperation of researchers from different disciplines such as information systems, business informatics, and computer science is required. BIR 2022 aimed to promote this collaboration to address the challenges above.

This year, the main conference track attracted 41 submissions with authors from 18 countries. Each submission was reviewed by at least four members of the Program Committee. As a result, 14 high-quality papers were selected for publication in this volume and for presentation at the conference. They cover different aspects of the conference's main topic as well as of business informatics research in general.

The conference program also included two keynotes: Mikael Lind, Professor at the Chalmers University of Technology (Sweden) and Senior Strategic Research Advisor at Research Institutes of Sweden (RISE), addressed "Maritime Informatics - a contributor for a high performing and sustainable maritime industry" and Phillip Wree, Head of Department for Smart Farming at Fraunhofer-Institute for Computer Graphics Research (Germany), gave a talk on "Smart Farming and Agricultural Digitization - with more Information for less Resource Demand".

We would like to thank everyone who contributed to the BIR 2022 conference. We thank the authors for contributing and presenting their research, we appreciate the invaluable contribution of the Program Committee members and the external reviewers, we thank the keynote speakers for their inspiring talks, we are grateful to the Springer team for providing excellent support regarding the proceedings production, and we thank all members of the local organization team from Rostock.

August 2022

Ērika Nazaruka
Kurt Sandkuhl
Ulf Seigerroth

# Organization

BIR 2022 was organized and hosted at the University of Rostock, Germany, during September 21–23, 2022.

## General Chair

Kurt Sandkuhl       University of Rostock, Germany

## Program Co-chairs

Ērika Nazaruka       Riga Technical University, Latvia
Ulf Seigerroth       Jönköping University, Sweden

## Doctoral Consortium Co-chairs

Björn Johansson       Linköping University, Sweden
Anne Gutschmidt       Rostock University, Germany

## Workshop Co-chairs

Janis Stirna       Stockholm University, Sweden
Hasan Koc       Berlin International University, Germany

## Local Organization Chairs

Jack Rittelmeyer       University of Rostock, Germany
Marcus Triller       University of Rostock, Germany

## Steering Committee

| | |
|---|---|
| Marite Kirikova (Chair) | Riga Technical University, Latvia |
| Björn Johansson (Vice Chair) | Lund University, Sweden |
| Eduard Babkin | National Research University Higher School of Economics, Russia |
| Robert Andrei Buchmann | Babeș-Bolyai University of Cluj Napoca, Romania |
| Rimantas Butleris | Kaunas University of Technology, Lithuania |
| Sven Carlsson | Lund University, Sweden |
| Peter Forbrig | University of Rostock, Germany |

| | |
|---|---|
| Dimitris Karagiannis | University of Vienna, Austria |
| Andrzej Kobyliński | Warsaw School of Economics, Poland |
| Raimundas Matulevicius | University of Tartu, Estonia |
| Lina Nemuraite | Kaunas Technical University, Lithuania |
| Jyrki Nummenmaa | University of Tampere, Finland |
| Małgorzata Pańkowska | University of Economics in Katowice, Poland |
| Andrea Polini | University of Camerino, Italy |
| Vaclav Repa | Prague University of Economics and Business, Czech Republic |
| Kurt Sandkuhl | University of Rostock, Germany |
| Benkt Wangler | University of Skövde, Sweden |

## Program Committee

| | |
|---|---|
| Gundars Alksnis | Riga Technical University, Latvia |
| Said Assar | Institut Mines-Telecom Business School, France |
| Eduard Babkin | National Research University Higher School of Economics, Russia |
| Per Backlund | University of Skövde, Sweden |
| Amelia Badica | University of Craiova, Romania |
| Peter Bellström | Karlstad University, Sweden |
| Erik Bergström | Jönköping University, Sweden |
| Catalin Boja | Bucharest Academy of Economic Studies, Romania |
| Dominik Bork | Technical University of Vienna, Austria |
| Tomáš Bruckner | Prague University of Economics and Business, Czech Republic |
| Robert Andrei Buchmann | Babeș-Bolyai University of Cluj Napoca, Romania |
| Sybren De Kinderen | University of Luxembourg, Luxembourg |
| Massimiliano de Leoni | University of Padua, Italy |
| Bruce Ferwerda | Jönköping University, Sweden |
| Hans-Georg Fill | University of Fribourg, Switzerland |
| Peter Forbrig | University of Rostock, Germany |
| Ana-Maria Ghiran | Babeș-Bolyai University of Cluj-Napoca, Romania |
| Jānis Grabis | Riga Technical University, Latvia |
| Knut Hinkelmann | FHNW University of Applied Sciences and Art Northwestern Switzerland, Switzerland |
| Adrian Iftene | "Alexandru Ioan Cuza" University of Iasi, Romania |
| Emilio Insfran | Universitat Politècnica de València, Spain |
| Amin Jalali | Stockholm University, Sweden |

| | |
|---|---|
| Florian Johannsen | University of Bremen, Germany |
| Björn Johansson | Linköping University, Sweden |
| Miranda Kajtazi | Lund University, Sweden |
| Dimitris Karagiannis | University of Vienna, Austria |
| Christina Keller | Lund University, Sweden |
| Marite Kirikova | Riga Technical University, Latvia |
| Markus Lahtinen | Lund University, Sweden |
| Michael Lang | National University of Ireland, Galway, Ireland |
| Michael Le Duc | Mälardalen University, Sweden |
| Moonkun Lee | Chonbuk National University, South Korea |
| Henrik Linderoth | Jönköping University, Sweden |
| Ginta Majore | Vidzeme University of Applied Sciences, Latvia |
| Raimundas Matulevicius | University of Tartu, Estonia |
| Andrea Morichetta | University of Camerino, Italy |
| Erika Nazaruka | Riga Technical University, Latvia |
| Jyrki Nummenmaa | Tampere University, Finland |
| Malgorzata Pankowska | University of Economics in Katowice, Poland |
| Victoria Paulsson | Linkoping, Sweden |
| Jens Myrup Pedersen | Aalborg University, Denmark |
| Dana Petcu | West University of Timisoara, Romania |
| Paul Pocatilu | Bucharest University of Economic Studies, Romania |
| Andrea Polini | University of Camerino, Italy |
| Dorina Rajanen | University of Oulu, Finland |
| Barbara Re | University of Camerino, Italy |
| Václav Řepa | Prague University of Economics and Business, Czech Republic |
| Ben Roelens | Open University of the Netherlands, The Netherlands |
| Christian Sacarea | Babeş-Bolyai University of Cluj-Napoca, Romania |
| Kurt Sandkuhl | University of Rostock, Germany |
| Ulf Seigerroth | Jönköping University, Sweden |
| Gheorghe Cosmin Silaghi | Babeş-Bolyai University of Cluj-Napoca, Romania |
| Jonas Sjöström | Uppsala University, Sweden |
| Janis Stirna | Stockholm University, Sweden |
| Ann Svensson | University West, Sweden |
| Torben Tambo | Aarhus University, Denmark |
| Gianluigi Viscusi | Imperial College Business School, UK |
| Anna Wingkvist | Linnaeus University, Sweden |
| Wiesław Wolny | University of Economics in Katowice, Poland |

| Jelena Zdravkovic | Stockholm University, Sweden |
| Alfred Zimmermann | Reutlingen University, Germany |

## Additional Reviewers

| Arianna Fedeli | University of Camerino, Italy |
| Amirhossein Gharaie | Linköping University, Sweden |
| Shahrzad Khayatbashi | Linköping University, Sweden |
| Morena Barboni | University of Camerino, Italy |
| Shashini Rajaguru | Linköping University, Sweden |

# Contents

# Information System Development

# Challenges of Low-Code/No-Code Software Development: A Literature Review

Karlis Rokis and Marite Kirikova(✉) ⓘ

Institute of Applied Computer Systems, Riga Technical University, Riga, Latvia
karlis.rokis@edu.rtu.lv, marite.kirikova@rtu.lv

**Abstract.** Low-code/no-code software development is an emerging approach delivering the opportunity to build software with a minimal need for manual coding and enhancing the involvement of non-programmers in software development. Low-code principles allow enterprises to save time and costs through a more rapid development pace and to improve software products quality by bringing closer together business and information technologies as well as promoting automation. Nevertheless, the low-code/no-code approach is a relatively new and continuously progressing domain that requires understanding of existing challenges and identification of improvement directions. In this paper, challenges in the low-code software development process and suggestions for their mitigation are identified and amalgamated with the purpose to deliver insights into the current state of the low-code/no-code development process and identify areas for further research and development.

**Keywords:** Low-code · No-code · Software development · Requirements · Low-code development platform · Citizen developer

## 1 Introduction

Low-code development is a software development approach that merges minimal hand-coding, graphical user interface, and visual abstraction. Development of applications is based on model-driven engineering principles, utilizing benefits delivered by cloud infrastructure (as development platforms are usually offered as Platform-as-a-Service), visual high-level abstraction, and automatic code generation [1, 2]. Practitioners use the term "no-code development" as a synonym to refer to low-code development practices [3]. Hereafter, terms low-code and no-code are not particularly distinguished and using term low-code also refers no-code and vice versa.

Low-code software development (LCSD) is an emerging trend as it proposes to address one of the main causes of delayed development – shortage of professional developers – by enabling non-programmers and non-technical personnel, called "citizen developers", to participate in the development process [1, 2]. Citizen developers are experts in a particular domain and system's functionalities meaning that they know and can define requirements. And LCSD through minimization of manual coding and by emphasizing visual platforms interfaces also enables participation of citizen developers in software development lifecycle phases – design, development, testing, deployment,

Ē. Nazaruka et al. (Eds.): BIR 2022, LNBIP 462, pp. 3–17, 2022.
https://doi.org/10.1007/978-3-031-16947-2_1

and maintenance. Although citizen developers, on the one hand, are considered as the main users of low-code development platforms (LCDP), they, on the other hand, lack programming knowledge and consequently this is a limiting factor for LCSD adaption and application by citizen developers [2, 4, 5].

Besides the advantage of enabling citizen developers to participate in the software development process, there are multiple other benefits as well. Low-code platforms enable the rapid translation of business requirements and development into the application as well as the possibility to quickly make the adjustments as there is no need for extensive manual coding. Additionally, due to reduced time spent on the development cycle, also development costs are reduced [3, 5]. However, several challenges in low-code software development should be considered and they are covered in this paper.

In the paper, a literature review has been conducted to amalgamate challenges in low-code software development and find, in the current literature, mitigation suggestions or directions regarding these challenges. The remainder of the paper is structured as follows: Sect. 2 includes the description of the research method. In Sect. 3, a background regarding low-code software development and related works is given. Section 4 covers identified challenges and mitigation suggestions followed by conclusions of the review in Sect. 5.

## 2  Methodology

A literature review is a helpful instrument to understand a currently existing body of knowledge in a specific domain, to identify the gaps and to see the direction in which further research is required [6]. The review has been conducted according to the approaches described in Levy *et al.* "A systems approach to conduct an effective literature review in support of information systems research" [6]. The goal of the paper is to review existing literature to identify challenges in the low-code/no-code software development and to find, in the current literature, suggestions, solutions or proposals on possibilities to mitigate and overcome these challenges. The following research questions were established to reach the defined goal:

**RQ1:** What challenges for low-code software development and implementation can be identified in the current literature?

**RQ2:** What solutions, actions or ideas could help to overcome these challenges?

The retrieved articles were validated against the selection criteria. The inclusion and exclusion criteria are described in Table 1.

Based on the topic of the paper and research questions, the keywords and search string were formed. Used keywords were "low code" and "no code" (joined with Boolean search connector "OR") and "develop*" and "implement*" (joined with operator OR).

**Table 1.** Selection criteria of the articles

| Selection | No | Criteria |
|---|---|---|
| Inclusion | I1 | The article covers the low-code/no-code software development |
| Exclusion | E1 | The article is not published in English |
| | E2 | The full text of the article is not accessible |
| | E3 | The article does not provide an answer to any of the research questions |

**Table 2.** Number of results per scientific library

| Database | Number of results | Primary studies |
|---|---|---|
| IEEE | 104 | 10 |
| ACM | 51 | 11 |
| ScienceDirect | 22 | 1 |

The automatic search was applied for the title, abstract and keywords fields of the papers. The final search string was formed:

("low code" OR "no code") AND (develop* OR implement*)

An initial keyword-based search was performed in the last quarter of 2021 in scientific databases: IEEE Xplore Digital Library, ACM Digital Library and ScienceDirect. These databases, as suggested in [7], are considered as relevant libraries in the software engineering domain. In the ACM Digital Library and ScienceDirect, the article type field was set to "Research Article". Additionally, the results not related to the domain of information technology (for instance, natural science, medicine, and other domains) were removed as they are not relevant for this review. In total 177 articles were found by the automatic keyword-based search. The number of found articles per scientific library is displayed in Table 2 column "Number of results". The results were reviewed by applying inclusion and exclusion criteria to the title and abstract. However, for a few articles to ensure they are relevant for the literature review, additionally the full text was reviewed. In the end, 22 articles were identified as primary studies relevant for the review based on keyword search.

In the article [6] it is emphasized that the keyword search is not sufficient for a literature search process, but it is just an initial step. As a second search strategy, the backward search was applied. Within this search, the references of the identified primary articles were reviewed. One additional study was identified through the backward search process.

In total 23 unique articles were selected for structured literature review. Gathered articles have been published from 2019 to 2021 showing that the most recent literature is included in the review.

## 3   Background and Related Work

In the following section, the general concepts that will be used further in the work are explained.

The development process of low-code/no-code applications can be divided into several phases. The initial and continuous step throughout the development process is *Requirements analysis*. The first phase – *Data modelling* – involves the configuration of the data schema for the application (entities creation, definition of relations, constraints, and dependencies). The following phase is a *Definition of a user interface (UI Design)*. This involves the creation of forms and pages that are applied for view definition of application. In the later stages, the definition of user access and security management for forms, pages, entities, and components is performed. For particular forms or pages, different workflows – business logic – might be needed and those are realised by implementing different operations for interface elements. These actions are performed in the third development phase called *Specification and implementation of business logic rules and workflows*. In the fourth phase, the *Integration of external services* is established. In such a way it is possible to consume external services directly in low-code software development by using application programming interfaces (APIs) provided by third parties. Then the *testing and application deployment* is performed. In the end, the *customer feedback and defined additional features* are gathered [1, 2].

Nowadays agile methodology is a widely used approach and its adoption is increasing [8]. Low-code/no-code software development process fosters the building of applications, gathering of users' feedback and implementing changes at a rapid pace delivering continuous increments and improving customer satisfaction. This matches the approach of agile development. Based on that, low-code software development stages can be related to the phases of the agile software development process [1]. To show how these stages refer to each other the article [1] proposes and presents the relation among them. A modified visualization of this relation is displayed in Fig. 1. The inner (yellow) circle groups the low-code software development stages, but the agile development stages are marked with the outer (light blue) area. Corresponding stages between methodologies are filled with the same pattern. In further work, where applicable, we will refer to the agile development stages as it is a commonly used methodology and low-code software development conforms well to agile principles such as continuous delivery and customer satisfaction. In addition, multiple LCDPs are equipped with facilities supporting agile development methodology [1, 2, 8, 9].

Low-code software development is accomplished on low-code software development platforms. These platforms are cloud- or on-premise-based and applications are developed through visual interfaces, the use of prebuilt components, their customization, and configurations of the settings. Using visual diagrams, high-level abstraction, declarative languages, and, in particular cases, also manual coding, the developers define

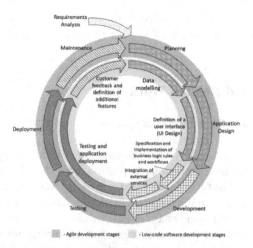

**Fig. 1.** Development stages in the agile and low-code software development process (adapted from the description in [1])

user interfaces, business logic, and data services [2, 4, 5]. The architecture of a low-code development platform is made of four main layers: (1) application layer, (2) service integration layer, (3) data integration layer, and (4) deployment layer [2].

The top, application, layer provides a graphical environment for users direct interaction to define the developed application, its user interfaces, and behaviour. This layer includes toolboxes and widgets for interface development as well as mechanisms for authentication and authorisation [2].

To utilise different services, for instance, using APIs, there is a dedicated layer distinguished for that called the Service integration layer [2].

The Data integration layer covers data integration issues allowing to operate and manipulate data coming from different sources [2].

The Deployment layer provides that the developed applications (depending on the applied platform) can be deployed either on cloud or on-premises environments [2].

If this architecture is expanded, a low-code platform consists of multiple components that can be arranged in three tiers. The first one is application modeller for applications specification using constructs and abstraction (including, for instance, widgets, connectors, business logic flows, data model, security rules, and others), the second tier forms server-side and its functionalities (for instance, platform server, compilers, optimizers, code generators, services), and the third tier is integrated external services (for instance, database servers, third-party systems, APIs, microservices, model repositories, and collaboration platforms) [2].

Low-code development has already been applied in multiple domains. The article [3] by an empirical study of forum posts, has identified, from practitioners' perspective, such application domains as e-commerce, business process management, social media, customer relationship management, extract, transform and load (ETL) processes, entertainment, content management systems, robotic process automation, and medicine. Also, in scientific literature, wide variety of application domains can be identified. In [9] Martins *et al.* develop a platform for human resource management; in [10], Oteyo *et al.* build the application in the domain of agriculture; Ihirwe *et al.*, in [11], present the state of research in the Internet of Things (IoT) domain; in [12], Waszkowski describes the application in business process automation in manufacturing; Arora *et al.*, in [13], identify such potential application domains as banking and communication; Varajão, in [14], shows the application in public health area; in [15], Kourouklidis *et al.* tell about application possibilities in machine learning model monitoring; Di Sipo *et al.* propose, in [16], implementation in the domain of recommender systems; in [17], Iyer *et al.* present an application in the geospatial domain; and Daniel *et al.*, in [18], present the development framework for chatbot implementation.

Some works review challenges regarding the low-code software development or application of low-code development platforms. Sahay *et al.*, in the technical survey "Supporting the understanding and comparison of low-code development platforms" (2020), by reviewing multiple low-code platforms and by developing benchmarking applications on them identify such challenges as interoperability, extensibility, and scalability issues as well as a need for extensive learning that is introduced by platforms characteristics [2]. A team led by Khorram worked on a paper discussing "Challenges & Opportunities in Low-Code Testing" (2020) comparing and analysing testing facilities in LCDP market leader platforms. Specific challenges, for instance, limited capabilities of the low-code testing framework, automated testing, and others were identified and grouped and opportunities for future work were delivered [4]. In the paper "An Empirical Study of Developer Discussions on Low-Code Software Development Challenges" (2021) by al Alamin *et al.*, by analysing the online developer forum *Stack Overflow* posts, identified the low-code software development challenges faced by practitioners using nine popular low-code development platforms. The paper revealed the most challenging topics, grouped them by their categories (whether a topic is related to customization, platforms adoption, database management or integration) and development stages proposing the most difficult low-code software development areas from a practitioners' perspective [1]. A similar empirical study "Characteristics and Challenges of Low-Code Development: The Practitioners' Perspective" was conducted by Luo Y. *et al.* who expanded analysis with another forum *Reddit* and the focus on development characteristics, benefits, and disadvantages in low-code development [3].

To the best of the authors' knowledge, currently, there is no literature review available that amalgamates challenges in low-code/no-code software development and possible ways or improvements that could help to overcome them in a stage-based manner. The findings presented further in this paper could be helpful for (i) practitioners to oversee challenges, and see how to meet and handle them, (ii) vendors to realise weaknesses that should be addressed and further development directions, and (iii) research community to identify areas that might require further research.

## 4   Challenges in Low-Code/No-Code Development

This section provides an answer to the research questions about the challenges of the low-code software development process and existing or possible suggestions, actions, solutions or ideas that could help to mitigate them. To structure the content of identified challenges from papers related to low-code/no-code development, the challenges are grouped according to software development phases described in Sect. 3. In Table 3, the identified challenges are presented according to these stages, and, next to them, also possible ways, found in the literature, that could help to overcome these challenges are recorded. Positioning of the text is made so that it would be visible which treatments refer to several challenges and which to just a specific challenge or a smaller group of challenges.

Further in this section, the challenges of each stage and their mitigation possibilities are discussed in separate dedicated subsections. LCSD challenges, which differ by the level of complexity, are not distributed evenly among development stages [1]. Considering count of identified challenges as well as their difficulty, more attention to challenges related to design and application development phases has been paid in the related work [1, 3].

### 4.1   Requirements Analysis

Requirements Analysis is the initial step to identify users' expectations regarding software to be developed [1]. The importance of requirements specification is high – requirements might influence the selection of the right platform and the validation (for instance,

**Table 3.** Challenges and mitigation possibilities

| Phase | Challenges | | Suggestions for mitigation | |
|---|---|---|---|---|
| | Name | Studies | | |
| Requirement analysis | Requirements specification | [1, 3, 4, 11] | • Development of minimum viable product [5, 14]<br>• Requirement management tools [1]<br>• A successful adaption of low-code technologies [5] | |
| | Changing requirements | [5, 14] | | |
| Planning | Selection of the platform | [10] | • Implementation of standards [2]<br>• List of features for platforms comparison [2, 10, 11, 19] | |
| | Vendor lock-in | [2–5, 18] | | |
| Application design | Extensibility limitations | [2, 18] | • Implementation of standards [2, 11] | |
| | Interoperability | [2, 11] | | |
| | Consideration of scalability | [2, 20] | • Further development of low-code development platforms [20] | |
| | UI design | [1] | • Address knowledge gap [1]<br>• Elaborated platforms' documentation [1] | |
| | Data storage design | [1] | | |
| Development | Implementation of business logic | [1, 3] | • Elaborated platforms' documentation [1]<br>• Learning resources [1–3] | |
| | Integration | [1, 3] | | |
| | No access to source code | [3] | • Further development of low-code development platforms [3, 20] | |
| | Customization of UI | [1, 3, 21, 22] | • Recommender systems for development [23] | • Automated conversion of design elements [21, 22] |
| | Debugging | [1, 3] | | • Solutions for debugging [1] |

*(continued)*

**Table 3.** (*continued*)

| Phase | Challenges | | Suggestions for mitigation | |
|-------|------------|---|----------------------------|---|
| | Name | Studies | | |
| Testing | Limited testing and analysis support | [1, 4, 11] | • Elaborated platforms' documentation [1]<br>• Introduction of general low-code testing framework [4, 11]<br>• Test automation [4, 13, 24, 25] | |
| | Dependence on third-party testing tools | [1, 4, 11] | | |
| | Testing of non-functional requirements | [4] | | |
| Deployment | Performance | [1, 3] | • Elaborated platforms' documentation [1] | |
| | Configuration issues | [1] | | |
| | Accessibility issues | [1] | | |
| | Version control | [1] | | • Use of repositories [1, 2] |
| Maintenance | Debugging | [1, 3] | • Elaborated platforms' documentation [1] | • Solutions for debugging [1] |
| | Use of maintenance features | [1, 19] | • Learning resources [1] | |

in a form of tests) whether the software meets users' expectations are performed based on specified requirements [3, 4]. The literature shows that support of requirements specification differs from one platform to another. For instance, the article [1], which focuses on the top nine low-code software development platform market leaders, states that platforms provide tools for requirement management, however [11] points out that requirements specification lacks the focus in most platforms dedicated for the domain of IoT. That means that requirements specification and management, depending on the used platform, can be challenging and practitioners would value having a requirements management tool in the low-code software development platform.

The changes of requirements might be considered as another challenging aspect mentioned in the literature within the requirements analysis phase as they have an impact on the software design [5]. However, successful application of the low-code principles which delivers an opportunity to develop at a rapid pace, handle this aspect well by enabling such exploratory requirements validation techniques as prototyping and building a minimum viable product. The low-code approach enables an opportunity to quickly develop a minimum viable product that can be validated against customer requirements to understand the value of it before putting an effort and resources into the development of functionality or features [5]. For instance, as described in [14], low-code principles allowed to quickly create initial functional prototype and to continuously deliver the functional solution the following day and to maintain the quality of the software artefacts even though requirements were constantly evolving.

## 4.2  Planning

The planning phase includes analysis and planning of feasibility, complexity, risk, dependencies, and timeline taking into consideration the operational aspects. The challenge identified in the literature in this phase is related to the selection of a suitable platform as there is the large number of platforms available in the market [2, 10]. The practitioners are concerned about related costs, the learning curve, supported platform features and functionality in development, deployment, and maintenance [1].

By analysing low-code platforms, the article [2] proposes 35 features that could help to compare and select an appropriate platform. Analysed features cover such issues as graphical user interfaces, support of interoperability, security, opportunities on collaborative development, reusability possibilities, scalability, mechanisms to specify business logic, application build mechanisms, deployment support, and what kind of application the development platforms support [2]. In addition to the features proposed by [2], multiple additional criteria for comparison have been included in other sources, for instance, features for specification of roles and users, support to requirements specification and verification, support of testing and validation, availability of artificial intelligence components [11, 19]. As noticed in [11], additional characteristics relevant to a particular domain can be promoted (for instance, as in the case of IoT, characteristic "focus on "The thing" layer" is essential for comparison). Application of such feature list and comparisons as proposed in [2, 11, 19] can be found in literature in practical examples. For instance, article [10], which covers the concern about the selection of the tool for the development of agriculture applications, uses similar criteria for making selection decisions. Also, the taxonomy for comparison of low-code development platforms provided by [2] has been put into the practice in experience report in [26] developing two scenarios on main market platforms. This shows that a feature list (taxonomy) for comparison and selection of platforms could be useful for practitioners as the basis for informed selection decisions [2].

Concerns regarding vendor lock-in which means that the platforms user is restricted and dependent on the provider, is one of the main reasons why companies are not adopting low-code platforms [5]. In article [18], the example of a chatbot platforms selection describes that companies end up in vendor lock-in by choosing a particular platform that is coupled to a particular engine, but such situations could be avoided by introducing channel and platform-independent frameworks. Vendor lock-in is related to such design aspects as extensibility and interoperability (see Subsect. 4.3 *Application Design* for detailed explanation) and could be mitigated by standardization, enhancing information exchange, and sharing of artefacts among different platforms [2].

## 4.3  Application Design

The specification of design which is based on applications requirements is determined in the application design phase. Design is considered regarding architecture, modularity, extensibility, scalability, and other aspects [1]. In the current literature, in this phase the following multiple low-code/no-code software development challenges are identified: limitations of extensibility and interoperability and considerations regarding scalability, data storage, and user interface design.

As identified in the literature, practitioners can expect that they will face extensibility limitations which refers to the possibility to refine or extend the provided functionalities by the tool, as new function addition to proprietary, closed source platforms is very difficult if not impossible [2, 18]. Also, interoperability – the tool's ability for internal (among components) and external (among services) information exchange – is a challenging aspect. Due to the lack of standards and reason that platforms are closed source, the interoperability possibilities are impeded causing limitations to architectural design and developed service and other artefact sharing. Both aspects contribute to concerns regarding vendor lock-ins making customers dependent on platform and vendors' decisions [2, 3]. To mitigate this challenge, it would be valuable to propose and apply the standards. The example given by [11] shows that it is possible to propose and introduce the standard (IoT reference model), which is adopted by different tools, even in such a complex domain as IoT. This example could serve as a starting point for further exploration and expansion of the standards covering interoperability issues [11]. Additionally, as stated under the *Planning* phase, it is relevant to consider aspects of extensibility and interoperability in platform selection criteria.

Another challenge in this phase is related to low-code development platforms scalability that covers platform ability to scale regarding the number of users, data traffic or data storage [2]. As explained in [20] users, regarding low-code platforms, respect their responsiveness and the ability to process complex procedures in a reasonable time. Therefore [20] proposes research lines for further development of low-code platforms to provide the multi-tenant environment for collaborative work, extend a model processing paradigm scaling possibility for a large number of elements and ensure engineers with a set of criteria for the selection of the appropriate model transformation engines in multi-tenant execution environments.

Practitioners have pointed out that during application design they face challenges regarding the design of user interfaces and data storage [1]. The user interface design process is complex, and it involves designing expected behaviour, "look and feel" of the system and it requires certain skills of the designer [22]. This seems to be challenging for one of the main low-code software platform user groups – citizen developers – as they might lack knowledge and design process experience [16]. Also, the challenge of data storage design, use of on-premises data sources, data migration to the platform, could be rooted in citizens developers' knowledge gap or information availability, for instance, in platforms' documentation, as platforms, in general terms, are designed so that they support different data storage configurations [1, 19]. The authors of this work could not identify a specific direction of actions in current literature that could help to mitigate these design challenges for citizen developers, however, the reduction of the knowledge gap, as well as the availability of elaborated documentation, can be suggested and might reduce the occurrences of these challenges.

## 4.4 Development

In this phase, the actual software is developed. Practitioners experience different kinds of challenges when customizing user interface, implementing business logic, integrating with third-party services, or solving more complex development issues that require access to code [1, 3]. For the topics related to user interface customization possibilities, implementation of business logic and integration, one of the identified reasons in the literature is incomplete or even incorrect documentation as well as the lack of learning resources, for instance, tutorials, of the platforms. Also, multiple platforms have non-intuitive interfaces, limited drag-and-drop capabilities and they require knowledge in software development, resulting in a high learning curve for the adoption of low-code technologies which is a challenging aspect for citizen developers [1–3]. However, the situation seems to be different in the case of developers with some coding experience. In the article [9], which describes the development of the low-code application for human resources self-service using the OutSystems[1] platform, the learning curve is identified as low for the team of developers, allowing to assume that the level of knowledge may impact the learning and adoption process of the technology.

In some cases, limited flexibility of the functionality and design of low-code development platforms arises the challenge of solving more complex development issues. If the platform does not provide enough customization features that may result in a need to write custom code spending more time, rising complexity, and requiring compromises on the product functionalities [3]. Additionally, challenges of third-party services integration depend on the platforms' extensibility capabilities which were discussed in Subsect. 4.3 *Application Design* [2].

Nevertheless, to mitigate previously stated development challenges from the learning and adaption perspective, platform vendors should provide elaborated documentation and learning resources. At the same time, practitioners should consider these challenges and realise possible trade-offs for low-code software development [1]. Another potential solution to assist citizen developers during the development stage is proposed in paper [23] by suggesting a recommender system that would use the captured knowledge from previously developed applications.

User interface customization challenges can be also viewed from another perspective in cases when user experience and user interface designers and the team of front-end developers collaborate in the process of converting user interface design elements and their customizations (designed using dedicated tools) into the low-code platforms' representation. This step requires manual processing which is not efficient and can end up in the inaccurate translation of customization, errors or gaps between the initial design and the result. An approach to mitigate this is the automated conversion of design artefacts into user interface components [21, 22]. Articles [22] and [21] described that such tools delivered improvements in terms of efficiency.

Another challenge that should be taken into consideration and accepted in low-code software development is difficulties related to debugging due to the graphical representation of software development kits in low-code development platforms [1, 3], however, current literature does not explain whether specific mechanisms or tools

---

[1] https://www.outsystems.com.

(for instance, debuggers) are applied although it is suggested for practitioners to adopt strategies to learn the process of debugging [1].

## 4.5 Testing

This phase includes tests on the developed system to verify that all the stated requirements are realized [4]. Multiple challenges are identified in the low-code software development testing phase. The practitioners report a lack of documentation, particularly, to run automated tests, test coverage, and the use of third-party testing tools [1]. These challenges could be mitigated by providing elaborated documentation.

To cover all testing activities, the largest low-code development platforms provide the opportunity for integration of third-party testing tools, however, due to dependency on these tools, there is limited involvement of citizen developers as they do not have the required level of knowledge. At the same time platforms have limited low-code testing frameworks and analysis support [4, 11]. Also, testing of non-functional requirements in low-code platforms is often neglected. Overall, these challenges could be addressed by researching and implementing a low-code testing framework that would cover all testing activities, would be reusable for other platforms, and be suitable also for citizen developers. Another suggestion for further research, concerning the low-code software development testing phase, is test automation. However, this automation due to the involvement of citizen developers should be provided with low technical knowledge dependency [4].

Regarding the support of testing and analysis activities, paper [24] presents a concept for executable tests generation for Process-Driven Applications to reduce manual test creation delivering time saving on this process and enabling citizen developers to generate test codes independently. Also, some platforms, for instance, Sagitec Software Studio S3[2], implementing Sagitec Test Studio, focus on reducing the creation of manual test cases and improving test executions and coverage [13]. Another article [25] reports on a prototype of a mocking mechanism based on the OutSystems platform to eliminate dependencies and to allow to test applications in isolation enhancing testing abilities of the platform. However, dependence on a specific platform as a downside can be mentioned for the last two proposals and it would be valuable to have such tools reusable for other platforms [4].

## 4.6 Deployment

Typically, low-code platforms integrate different components for development and among them also deployment assistants to perform the deployment easily and successfully [1, 19]. Nevertheless, practitioners share challenges regarding issues of deployment configuration, accessibility, and performance (slow loading and publishing) [1, 3]. In the current literature, particular mitigation opportunities to address these challenges are not provided, however, paper [1], through the example of deployment issues, points out the incomplete platform documentation that should be improved. Also, version control is identified by practitioners as a challenge, although platforms, for this purpose, provide repositories for storing artefacts and handling version control tasks [1, 2].

---

[2] https://www.sagitec.com/resource-development-tools-s3.

### 4.7 Maintenance

After the application is released, it needs maintenance support, to rapidly implement changes for the developed solutions or features to provide continuous availability. Low-code platforms are characterized as easy maintainable [1, 5]. In this phase debugging appears as a challenge (see Subsect. 4.4 *Development*) [1, 3]. Additionally, paper [1] indicates challenges regarding such platforms' maintenance features as monitoring, collaboration, and others, for instance, such as machine status and anomalies monitoring dashboards in ADAMOS platform[3] [5] and roles management. Reviewing previously identified challenges and suggestions in the paper [1], we can conclude that also for these challenges elaborated documentation and information availability could help to some degree to overcome them.

## 5 Conclusion

The objective of this literature review was to survey existing literature to identify challenges in the low-code/no-code software development and to find suggestions, solutions, and proposals on possibilities to mitigate and overcome these challenges. The review included 23 sources used to identify challenges and suggestions for their mitigation. The obtained results were organized according to seven agile software development phases.

Low-code technologies are a recently emerging trend and, even though low-code software development provides multiple benefits, there are plenty of challenges and improvements opportunities.

The literature review shows that low-code/no-code software development and technologies have several challenging aspects that call for further research and improvements, for instance methods and approaches for supporting citizen developers, minimal viable product based requirements management, and frameworks for standardization.

Multiple challenges to some degree could be mitigated by vendors by improving platforms' documentation and learning resources. Vendors should continue to work on platforms development so to address challenges and limitations identified in the design (e.g., extensibility, interoperability, scalability, and others) and development phases (customization and implementation options, integration possibilities, and others) to meet users expectations. Attention should be paid to improving the design and customization of the user interface as any custom adjustments which are not directly supported by the low-code platform may rise challenging aspects for developers or require compromising the functionality of the application.

Current review of challenges and their mitigation opportunities in low-code development suggests that practitioners should identify and address the knowledge gap, evaluate and analyse the low-code technologies adoption, and explore its opportunities to release its potential.

The whole low-code community that includes researchers, vendors and practitioners may benefit from joint efforts in considering the implementation of standards, introducing common frameworks (for instance, for testing) and developing new tools to enhance further progress of low-code technologies.

---

[3] https://www.adamos.com/en/.

# References

1. al Alamin, M.A, et al: An empirical study of developer discussions on low-code software development challenges. In: Proceedings – 2021 IEEE/ACM 18th International Conference on Mining Software Repositories, MSR 2021. Institute of Electrical and Electronics Engineers Inc., pp. 46–57 (2021)
2. Sahay, A., Indamutsa, A., di Ruscio, D., Pierantonio, A.: Supporting the understanding and comparison of low-code development platforms. In: Proceedings – 46th Euromicro Conference on Software Engineering and Advanced Applications, SEAA 2020. Institute of Electrical and Electronics Engineers Inc., pp. 171–178 (2020)
3. Luo, Y., et al: Characteristics and challenges of low-code development: the practitioners perspective. In: International Symposium on Empirical Software Engineering and Measurement. IEEE Computer Society (2021)
4. Khorram, F., Mottu, J., M., Sunyé, G.: Challenges & opportunities in low-code testing. In: Proceedings – 23rd ACM/IEEE International Conference on Model Driven Engineering Languages and Systems, MODELS-C 2020 – Companion Proceedings, pp. 490–499. Association for Computing Machinery Inc (2020)
5. Sanchis, R., García-Perales, Ó., Fraile, F., Poler, R.: Low-code as enabler of digital transformation in manufacturing industry. Appl. Sci. (Switzerland) **10**, 1–7 (2020). https://doi.org/10.3390/app10010012
6. Levy, Y., Ellis, T., J.: A systems approach to conduct an effective literature review in support of information systems research. Inf. Sci. **9**, 181–211 (2006). https://doi.org/10.28945/479
7. Kitchenham, B., Charters, S.: Guidelines for performing Systematic Literature Reviews in Software Engineering. Keele, Staffs (2007)
8. Fitriani, W., R., Rahayu, P., Sensuse, D., I.: Challenges in agile software development: a systematic literature review. In: 2016 International Conference on Advanced Computer Science and Information Systems, ICACSIS 2016, pp. 155–164. Institute of Electrical and Electronics Engineers Inc. (2017)
9. Martins, R., et al: An overview on how to develop a low-code application using OutSystems. In: Proceedings of the International Conference on Smart Technologies in Computing, Electrical and Electronics (ICSTCEE 2020), 9–10 October 2020, Virtual Conference (2020)
10. Oteyo, I.N., et al.: Building smart agriculture applications using low-code tools: the case for discopar. In: IEEE AFRICON Conference. Institute of Electrical and Electronics Engineers Inc (2021)
11. Ihirwe, F., et al.: Low-code engineering for internet of things: a state of research. In: Proceedings – 23rd ACM/IEEE International Conference on Model Driven Engineering Languages and Systems, MODELS-C 2020 – Companion Proceedings, pp. 522–529. Association for Computing Machinery Inc. (2020)
12. Waszkowski, R.: Low-code platform for automating business processes in manufacturing. In: IFAC-PapersOnLine, pp. 376–381. Elsevier B.V. (2019)
13. Arora, R., Ghosh, N., Mondal, T.: Sagitec software studio (S3) – a low code application development platform. In: 2020 International Conference on Industry 4.0 Technology, I4Tech 2020, pp. 13–17. Institute of Electrical and Electronics Engineers Inc. (2020)
14. Varajão, J.: Software development in disruptive times. Queue **19**, 94–103 (2021). https://doi.org/10.1145/3454122.3458743
15. Kourouklidis, P., Kolovos, D., Matragkas, N., Noppen, J.: Towards a low-code solution for monitoring machine learning model performance. In: Proceedings – 23rd ACM/IEEE International Conference on Model Driven Engineering Languages and Systems, MODELS-C 2020 – Companion Proceedings, pp. 423–430. Association for Computing Machinery Inc. (2020)

16. di Sipio, C., di Ruscio, D., Nguyen, P.T.: Democratizing the development of recommender systems by means of low-code platforms. In: Proceedings – 23rd ACM/IEEE International Conference on Model Driven Engineering Languages and Systems, MODELS-C 2020 – Companion Proceedings, pp. 471–479. Association for Computing Machinery, Inc. (2020)

17. Iyer, C.V.K., et al.: Trinity: a no-code AI platform for complex spatial datasets. In: Proceedings of the 4th ACM SIGSPATIAL International Workshop on AI for Geographic Knowledge Discovery, GeoAI 2021, pp. 33–42. Association for Computing Machinery Inc. (2021)

18. Daniel, G., Cabot, J., Deruelle, L., Derras, M.: Xatkit: a multimodal low-code chatbot development framework. IEEE Access **8**, 15332–15346 (2020). https://doi.org/10.1109/aCCESS.2020.2966919

19. Bock, A.C., Frank, U.: In search of the essence of low-code: an exploratory study of seven development platforms. In: 2021 ACM/IEEE International Conference on Model Driven Engineering Languages and Systems Companion (MODELS-C), pp. 57–66. IEEE (2021)

20. Horváth, B., Horváth, Á., Wimmer, M.: Towards the next generation of reactive model transformations on low-code platforms: three research lines. In: Proceedings – 23rd ACM/IEEE International Conference on Model Driven Engineering Languages and Systems, MODELS-C 2020 – Companion Proceedings. Association for Computing Machinery Inc., pp. 441–450 (2020)

21. Bexiga, M., Garbatov, S., Seco, J.C.: Closing the gap between designers and developers in a low code ecosystem. In: Proceedings – 23rd ACM/IEEE International Conference on Model Driven Engineering Languages and Systems, MODELS-C 2020 – Companion Proceedings, pp. 413–422. Association for Computing Machinery Inc. (2020)

22. Pacheco, J., Garbatov, S., Goulao, M.: Improving collaboration efficiency between UX/UI designers and developers in a low-code platform. In: 2021 ACM/IEEE International Conference on Model Driven Engineering Languages and Systems Companion (MODELS-C), pp. 138–147. IEEE (2021)

23. Almonte, L., Cantador, I., Guerra, E., de Lara, J.: Towards automating the construction of recommender systems for low-code development platforms. In: Proceedings – 23rd ACM/IEEE International Conference on Model Driven Engineering Languages and Systems, MODELS-C 2020 – Companion Proceedings, pp. 451–460. Association for Computing Machinery Inc. (2020)

24. Schneid, K., Stapper, L., Thone, S., Kuchen, H.: Automated regression tests: a no-code approach for BPMN-based process-driven applications. In: Institute of Electrical and Electronics Engineers (IEEE), pp. 31–40 (2021)

25. Jacinto, A., Lourenço, M., Ferreira, C.: Test mocks for low-code applications built with OutSystems. In: Proceedings – 23rd ACM/IEEE International Conference on Model Driven Engineering Languages and Systems, MODELS-C 2020 – Companion Proceedings, pp. 530–534. Association for Computing Machinery Inc. (2020)

26. Gurcan, F., Taentzer, G.: Using Microsoft PowerApps, Mendix and OutSystems in two development scenarios: an experience report. In: 2021 ACM/IEEE International Conference on Model Driven Engineering Languages and Systems Companion (MODELS-C), pp. 67–72. IEEE (2021)

# Ex-Post Evaluation of Data-Driven Decisions: Conceptualizing Design Objectives

Nada Elgendy[1]([⊠]) [iD], Ahmed Elragal[2] [iD], Markku Ohenoja[3] [iD],
and Tero Päivärinta[1] [iD]

[1] M3S, Faculty of Information Technology and Electrical Engineering, University of Oulu, Oulu, Finland
nada.sanad@oulu.fi

[2] Department of Computer Science, Electrical and Space Engineering, Lulea University of Technology, Lulea, Sweden

[3] Environmental and Chemical Engineering, Faculty of Technology, University of Oulu, Oulu, Finland

**Abstract.** This paper addresses a need for developing ex-post evaluation for data-driven decisions resulting from collaboration between humans and machines. As a first step of a design science project, we propose four design objectives for an ex-post evaluation solution, from the perspectives of both theory (concepts from the literature) and practice (through a case of industrial production planning): (1) incorporate multi-faceted decision evaluation criteria across the levels of environment, organization, and decision itself and (2) acknowledge temporal requirements of the decision contexts at hand, (3) define applicable mode(s) of collaboration between humans and machines to pursue collaborative rationality, and (4) enable a (potentially automated) feedback loop for learning from the (discrete or continuous) evaluations of past decisions. The design objectives contribute by supporting the development of solutions for the observed lack of ex-post methods for evaluating data-driven decisions to enhance human-machine collaboration in decision making. Our future research involves design and implementation efforts through on-going industry-academia cooperation.

**Keywords:** Data-driven decisions · Ex-post evaluation · Design objectives · Collaborative rationality · Human-machine collaboration

## 1 Introduction

The ex-post evaluation of data-driven decisions emerges as an increasingly relevant, whilst difficult, topic [7–9]. Its complexity lies in that data-driven decision making involves five interrelated elements: the human decision maker, machine (analytics algorithms), data, decision-making process, and decision outcome [9]. This coexistence of machine learning (ML) and artificial intelligence (AI) systems with human decision makers has ignited interest in augmenting human intelligence and capabilities, resulting in more "intelligent" data analysis and support for decision-making and learning [13,

Ē. Nazaruka et al. (Eds.): BIR 2022, LNBIP 462, pp. 18–34, 2022.
https://doi.org/10.1007/978-3-031-16947-2_2

18, 42]. However, ex-post evaluation is crucial for enabling feedback for experiential learning to improve both organization and machine decisions, and measure the benefits of human-machine collaboration [20, 24, 33, 42]. It can result in organizational and experiential learning [2, 23], rationalization [22], and sensemaking [51] from the decision outcomes and consequences, as well as allow for analysis, benchmarking, and comparison of the results [24]. Understanding how, why, and to what extent ML and AI systems are being used in decision making and their influence on individual and organizational decisions [8] is difficult to assess without a holistic evaluation perspective. This requires a feedback loop between actions and outcomes, and encoding the past into rules and procedures for future learning [23].

Nevertheless, despite its importance as one of the main stages in classical decision-making processes, ex-post evaluation is commonly overlooked in recent data-driven decision making research. Decision evaluation is more complicated than the mere evaluation of a choice [52]. Data-driven decision evaluation is further complicated by the fact that many interrelated factors and metrics affect the evaluation, involving both humans and machines in a constantly changing environment.

Shrestha et al. [36] distinguish further between three categories of human/machine decisions:

1) purely machine decisions (e.g., recommender systems, personalized ads);
2) sequence-based decisions, which can involve two sub-types:

   (a) human-to-machine (e.g., sports analytics on which the human expert seeks evidence from data);
   (b) machine-to-human (e.g., ideation in innovation);

3) aggregated decisions involving both humans and machines, in peer-like group decision making (e.g., assisted medicine healthcare applications).

This research focuses on the last two categories, in line with Ransbotham et al.'s [33] modes of collaboration where AI recommends and the human decides, or AI generates insights which the human uses in the decision process, or the human generates hypothetical situations and relies on AI to evaluate and assess them.

Such collaboration results in decisions provided by machines, decisions made by humans, and the final data-driven decision which is selected, implemented, and leads to certain outcomes, which all require evaluation. For example, let us consider an AI system used in clinical decision support. Depending on the context of the decision, one performance measure might be more important than another, such as with predicting mortality which requires high accuracy and precision [21]. Nevertheless, erroneous diagnoses can be made based on differences in the training data, and ground truth labels may not always be correct and are subjective to different opinions [19, 21]. Thus, a highly accurate model based on the available training data cannot indicate an accurate or correct decision, nor positive outcomes. Human intervention and monitoring are necessary, yet if their diagnosis conflicts with that of the machine, which one is correct? Accordingly, some outcomes (e.g., correct diagnosis) can only be known after time in order to evaluate whether or not the data-driven decision was, in fact, accurate and better than purely human decisions, which necessitates ex-post evaluation.

Due to a lack of information about how to perform such evaluation and the lack of IT artifacts to employ, organizations rarely conduct ex-post evaluation of their past decisions. Ex-post evaluation is designed to help companies learn from their mistakes in the past, avoid repeating them in the future, and convey their knowledge to others. It does not impede decision-making or promote a no-decision scenario, because the goal is to learn from the decision and improve the quality of future decisions, not to evaluate the decision-maker (human or machine).

Furthermore, a comprehensive viewpoint to the multiple, socio-technical, elements involved in data-driven decision making is lacking [25], and there is little agreement in the literature on what and how to evaluate [17, 41]. The aspect of time and the requirements for a longitudinal, or processual evaluation remains ignored, which would be imperative to capture the complex dynamics involving change related to multi-faceted decisions [5]. Accordingly, we set out with the following research question:

*"What are the requirements and design objectives for ex-post evaluation of data-driven decisions in organizations?"*

This research problematizes ex-post evaluation of data-driven decision making by highlighting the gap in research and the industry need for a more holistic solution. We define design objectives (DOs) for the solution by first extracting the relevant ex-post evaluation concepts from the literature. These concepts are then exemplified through an industrial example of a chemical production plant to foresee how ex-post evaluation of data-driven decisions could be done in practice, and accordingly outline the initial requirements for a design solution.– covering two first steps of a design research program (cf. [29]) with industry.

The remainder of the paper is structured as follows. Section 2 covers the related research and literature analysis. The research method is outlined in Sect. 3. Section 4 describes the results and finding, which are the evaluation requirements and DOs. Finally, Sect. 6 concludes the paper with suggestions for further research.

## 2 Literature on Evaluating Data-Driven Decisions

### 2.1 Lack of Ex-Post Evaluation Support

In search of evaluation concepts, criteria, or solutions, we reviewed literature from various streams and disciplines, including decision research, information systems (IS), behavioral sciences, AI, ML, and information technology (IT). In the following, the literature was divided roughly into three streams.

The first stream focuses on decision theories with attention on human rationality [10, 22, 37, 47] and decision making [1, 11, 27] in various fields, such as management, economics, and psychology. In this stream, ex-ante evaluation of alternatives and choices was extensively studied, often focusing on individual metrics and values (e.g., utility). Although ex-post evaluation was included as a stage of suggested decision processes and deemed crucial in some fields (e.g., policy making [50]), less attention was paid on how evaluation was conducted (methods, metrics, time) or how it influenced the remainder of the process. The collaboration between humans and machines and how to evaluate data-driven decisions was non-prevalent in this stream.

The second stream focuses on AI and ML from the technical perspective of computer science and engineering, and the application of algorithms, models, and methods to a dataset to solve a specific problem [18, 35]. Research tends to focus on the use of machines for selecting among ex-ante alternatives [25, 45]. Evaluation covers the performance of the algorithm or model used in decision making, and a limited set of evaluation metrics are prevalent in the field [26, 49]. Different metrics have their strengths and limitations, are dependent on the available data, and may be conflicting (efficiency vs. accuracy vs. cost, etc.) [34]. Moreover, evaluating model performance is not the same as evaluating the resulting decision or its consequences.

The third stream of research involves data-driven decision making, highlighting the sociotechnical aspect and the relationship between the human and machine decision makers, mainly in an organizational context and with an IS perspective [9, 20, 36, 46]. Evaluation is still generally limited to the evaluation of choices and the evaluation of the performance of algorithms and models. Limited sources considered evaluating outcomes [43], let alone with multiple metrics [14]. However, no holistic evaluation solutions were found to consider the data-driven decision as a whole.

Consequently, the interaction between humans and machines and their roles in decision making is still not clear, and further research is necessary to evaluate the resulting decisions and determine the benefit, impact, and learning achieved through human-machine collaboration. Hence, we need new ways to evaluate AI-enabled decisions and benefits of human-machine collaboration in data-driven decision making. [7–9, 20].

## 2.2 Ex-Post Evaluation Concepts for Data-Driven Decisions

The literature introduces various concepts relevant to evaluating data-driven decisions. These serve as a theoretical basis for the requirements analysis leading to the suggested DOs (cf. The left column of Table 2 in Sect. 4.1). First, there are embedded *contexts* for examining the decision situation to comprehend the factors affecting the decision and its impact [32]. The context pertains to the types of decisions made at different levels, ranging from individual to global, with varying requirements, as well as the decision environment, both internal and external [24]. The *environmental context* is the broadest perspective and includes the external environment and circumstances. The *organizational context* covers the characteristics of the organization in which the decision was made. The *decision context* includes aspects regarding the focus of the decision and the reasons for it, its relationship to other decisions, the complexity of the decision, constraints, etc. [24, 30, 32, 39].

*Time* highlights the processual nature of evaluation and refers to when and how often the evaluation is conducted, since the outcomes of the decision may vary across time. Decisions should be viewed from the perspective of process science which is concerned with understanding processes and influencing change in the desired directions over time [5]. One of the core requirements is to understand the emergent, situational, and holistic features of the decision, or the decision-making process, in its changing context [30], which adds to the necessity of a multi-faceted, process-oriented decision evaluation.

Data-driven decisions comprise *data-driven decision elements*, which include the *decision maker*, the *decision-making process*, the *data*, the *analytics/machine*, and the *decision outcome* [9]. Nevertheless, identifying decision outcomes is a difficult challenge due to their multi-faceted nature, variability in interpretation, acceptability and accountability to stakeholders, volatility and change, as well as their difficulty to fully grasp or quantitatively measure through indicators of success [27].

Accordingly, the decision outcome may be evaluated through multiple concepts, which extend across the various contexts and vary with time. Of particular importance are the *impact and consequences* of the decision and its perceived gains and losses [4, 27, 48, 52]. Furthermore, there is the *conformance* of the decision to certain criteria, as a decision involves some goals or values, some facts about the environment, and some inferences drawn from the values and facts. It must comply with objectives, criteria, standards, rules, and regulations, not only at the organizational context, but also at the environmental and societal contexts, since the decision may impact each [4, 27, 52].

Various *metrics* can be used for evaluating data-driven decisions and decision alternatives. Data-driven decisions are often evaluated with over-reliance or unwarranted dependence upon quantification and quantitative data [32]. Such metrics may potentially be conflicting, and generally focus on evaluating decision alternatives which differs from the ex-post evaluation of the decision after it is made [52].

*Errors and biases* can affect the outcome of decisions and thus need to be pinpointed and evaluated. Algorithmic predictions, although susceptible to their own types of errors, may influence human decisions. It is also necessary to differentiate between errors and biases that stem from the decision maker, and those which stem from the data, analytics, or machine, since each should be managed differently and require pertinent action [31].

# 3 Research Method

## 3.1 Research Design and Process

This research covers the first two steps of a design science research (DSR) process, to identify and motivate the problem, and to define the objectives of a solution [28]. Artifacts and solutions should be based on the relevant business needs from the environment and the applicable knowledge gained from the knowledge base [15]. Accordingly, the relevant concepts for ex-post evaluation were extracted from the literature (Sect. 2.2). These concepts were used to theoretically support the industrial case and categorize the interview questions and thematize the evaluation requirements (Sect. 4.1).

For portraying the practical aspects of our research, a case example of a chemical production plant is utilized. This plant, named ChemML (anonymized), is a simplified abstraction of a larger organization collaborating in an ongoing project, enhanced by the extensive knowledge and expertise of one of the authors experienced in chemical process engineering and decision making in such processes, and knowledgeable of the decisions, roles, data, and processes of ChemML and other chemical production plants.

This example was selected as chemical production plants have a high availability of mission-critical data-driven decisions. In such processes, hundreds or thousands of sensors routinely measure and automatically record data with high frequency. In a short time period, massive volumes of data are collected for process monitoring, evaluation, and control, which requires transforming the data into information for business and operation decision-making, in which ML tools have an important role [6, 44]. Furthermore, ChemML has recurring, operational, data-driven decisions in its production process, long-lasting adoption of systems utilizing ML in supporting decision making, and a desire to further evaluate, automate, and enhance the data-driven decision making processes.

Two expert interviews were held, from viewpoints of both the production planner and process operator roles, to discuss ChemML's data-driven decisions, explore the current evaluation methods, and discuss the need and requirements for a desired evaluation solution. By comparing the current and desired approaches for decision evaluation stated in the interviews, and applying deductive thematic analysis [38], we summarized the results into a set of evaluation requirements for each of the concepts. According to their functional similarities, the requirements were further thematized and mapped to more abstract and implementable DOs for an evaluation solution. The requirements were revised again to ensure that each requirement mapped to at least one DO.

The value of this study resides in the Eval 1 stage of Sonnenberg and vom Brocke's [40] DSR evaluation process for designing artifacts. This initial evaluation is conducted to justify a solution's novelty and importance for practice and to ensure that a meaningful problem has been identified. Accordingly, we attempted to evaluate feasibility, understandability, simplicity, completeness, and level of detail of the evaluation concepts and DOs in future design of an ex-post evaluation solution.

Internal validity was achieved through revision and agreement on the evaluation concepts, requirements, and DOs by each of the authors and expert in the case. The evaluation concepts and DOs were further presented to, and validated by, four experts in external software organization, under a case for utilizing data-driven decision making and AI to predict and prevent customer churn. The interview questions were validated by one of the analytics experts in the organization, and an additional interview was conducted with a customer success expert. The results were found to support the case of ChemML and findings of this paper, thus supporting external validity.

### 3.2 Case of ChemML

The production planning problem is a typical example of a complex, data-driven decision process with many interrelated factors, constraints, and major impacts. The orders placed by customers put a great pressure on production. Adjusting the production sequence must be done carefully to avoid disruption of production cycles, such as reduction in production rate and shutdowns. Moreover, ramping up the process and recovering from interruptions requires expenditure of energy, thus increasing the environmental load of the plant. Abrupt product sequence changes may cause quality deviations and wear of the machines and equipment.

Figure 1 (a) depicts a simplified flowsheet of ChemML's multi-step, multi-product production process. Two critical features complicate the decision making:

(1) production planning is based on make-to-order (MTO) as the production batches cannot be stored for prolonged times, and

(2)  routine operation has slow feedback from product quality to operational decisions.

Figure 1 (b) illustrates data flows of ChemML. The automation system data includes the sensory measurements such as temperature (TI), material consumption (FI), quality attributes (QI), and energy consumption (EI) from the production process. ML tools infer data to routine operations and predict performance for manual production planning. The data-driven tools are advisory since the final decisions (process operation and resource planning) need to be made by humans due to responsibility issues.

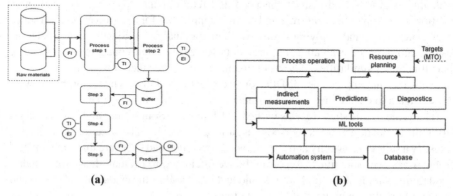

**Fig. 1.** **(a)** Production process schematic; **(b)** Data flows and decision support architecture

Based on the interviews, the data-driven decisions from the viewpoints of both the production planner and process operator roles, as well as the need for ex-post evaluation are described below in Table 1 (due to confidentiality requirements, some details could not be disclosed).

**Table 1.**  Data-driven decisions at ChemML

| Decision context | Production planner viewpoint | Process operator viewpoint |
|---|---|---|
| Description | • Determine and plan the production targets and capacities for a specific time interval and schedule the production process (weekly) | • Operation decisions (continuous) during the execution of the process. Includes choosing set points for the process (such as feed rates and temperature), steering the process, and avoiding/overcoming fault situations |
| Purpose | • Optimize production rate and product portfolio to meet market demand | • Optimize the process in terms of efficiency (energy, material), avoid faults, and solve possible problems |

(continued)

**Table 1.** (*continued*)

| Decision context | Production planner viewpoint | Process operator viewpoint |
|---|---|---|
| Decision Maker(s) | • Production planner (human decision maker) determines objectives and constraints<br>• ML tool supports decision by simulating scenarios and suggesting alternative schedules<br>• Human selects best schedule to meet designated criteria and makes the final decision<br>• Human may overlook output of the ML tool and decide not to use it | • ML tool provides outputs, insights, and predictions based on data and process parameters to steer the process and dynamically overcome fault situations<br>• ML tool provides suggestions of values for optimizing process efficiency<br>• Final decisions are made by the process operator who may use their own knowledge and expertise, along with additional monitoring methods |
| Mode of collaboration | • AI recommends, human decides | • AI recommends, human decides<br>• AI generates insights, human uses in decision process |
| Additional requirements (environment, organization) | • Meet sales demands and maximize profit<br>• Conform to safety and quality requirements, meet standards and regulations, and laboratory testing<br>• Minimize waste and carbon footprint, and conform to pollution limits and the use of hazardous materials | • Meet production targets in time<br>• Conform to safety and quality requirements, and professional standards and regulations |
| Need for ex-post evaluation | • Assess reliability and effectiveness of ML tool<br>• Enhance, both human and machine, learning from evaluation feedback<br>• Evaluate the collaboration between the human and machine<br>• Evaluating decisions at different time intervals would give indications if the reliability of the ML tool is increasing across time | • Evaluate ML tool and its value to decision making<br>• Evaluate ML indicators and their usefulness in decision making<br>• Evaluate the extent to which ML tool is used and affects the decision<br>• Evaluate the decision outcome<br>• Evaluate expertise of the process operator, and the monitoring methods used to reach the decision<br>• Evaluate uncertainties in measurement data and their effect on the decision |

# 4   Results and Findings

## 4.1   Analysis of Ex-Post Evaluation Requirements

Table 1 summarizes the requirements and considerations found necessary in the case for ex-post evaluation, in light of the analytical concepts originating in our literature review. Each of these conceptual elements helped to identify the interrelated requirements. Accordingly, we thematically grouped the similar requirements together from which we derived four main DOs, explained below. Each requirement in the table is labelled with the pertaining DO it corresponds to.

**Table 2.** Data-driven decision evaluation requirements

| Evaluation concepts | Proposed evaluation requirements/considerations |
| --- | --- |
| Overall evaluation | Define the evaluation metrics. Evaluation should not revolve purely around a single metric. (*DO1*) |
| | Determine the evaluation process, the evaluators, and their roles. (*DO1, DO2, DO4*) |
| | Differentiate between the evaluation of the ML tool output/decision, and the evaluation of the overall decision involving both humans and machines. (*DO1, DO2, DO3*) |
| | Simplicity of evaluation, without requiring much time or work. (*DO1, DO2, DO4*) |
| | Transparency of the evaluation process to increase trust in the ML tools. (*DO1, DO2, DO3, DO4*) |
| | Identify relevant criteria for each data-driven decision evaluation. Some criteria and elements need only be evaluated when triggered by change, or problems arise. (*DO1, DO2, DO4*) |
| | Automated/partially automated evaluation. (*DO1, DO2, DO4*) |
| | Continuous feedback and learning from past decisions. (*DO2, DO4*) |
| Evaluation across contexts (Decision, organization, environment) | Determine the relevant criteria and metrics in each of the contextual levels. (*DO1, DO2*) |
| | Determine the interrelationship between the metrics across the contextual levels, and how they affect the data-driven decision and its evaluation. (*DO1, DO4*) |
| | Determine what is being evaluated (decision/set of decisions) and by whom. (*DO1, DO2, DO4*) |
| Evaluation across time (Processual) | Determine the time intervals and periods for which certain types of decisions on various levels should be evaluated and/or re-evaluated. (*DO1, DO2, DO4*) |
| | Account for changes in decision related concepts and contexts (*DO1, DO2, DO4* |

(continued)

**Table 2.** (*continued*)

| Evaluation concepts | | Proposed evaluation requirements/considerations |
|---|---|---|
| Data-driven decision elements | Decision maker | Include both the human and the machine decision makers and suggest various metrics or criteria for evaluating each type of decision maker. (***DO1, DO3***) <br> Enhance learning of the decision makers based on the results of past decisions. (***DO3, DO4***) <br> Differentiate evaluation according to the mode of collaboration between the human and the machine. This may call for different evaluation methods, metrics, and requirements for different modes of collaboration. (***DO1, DO2, DO3, DO4***) <br> Evaluate decision maker-related aspects; it may be useful in learning from past decisions. (***DO3, DO4***) |
| | Process | In this case not required, or too difficult to evaluate. (***DO1, DO2***)) |
| | Data | Determine criteria and metrics for evaluating the data. (***DO1***) <br> Suggest the effect of the data or changes in the data, on the decision. (***DO1, DO2***) <br> Provide simple, automated methods for evaluating the data. (***DO1, DO4***) <br> Distinguish between the data required for making the decision, and the data required for evaluating the decision. (***DO1, DO2***) |
| | Analytics/ machine | Incorporate additional metrics and deal with conflicting metrics. (***DO1***) <br> Suggest the effect of the analytics (and choice of analytics) on the decision, and how to evaluate and incorporate the ML output. (***DO1, DO2, DO3, DO4***) <br> Enhance learning of the ML tool and feed the results back into the training data. (***DO3, DO4***) |
| | Decision outcome | Determine metrics and criteria for evaluating the outcome of the decision after it is made (what defines a "good" decision?). (***DO1, DO2***) <br> Observe the effect of the other decision factors, and their changes, on the decision outcome. (***DO1, DO2, DO4***) <br> Determine when the decision should be evaluated. (***DO1, DO2, DO4***) <br> Consider changing outcomes and the temporal factor. (***DO1, DO2, DO4***) |
| Impact and consequences | | Determine the metrics and criteria for evaluating the impact and consequences of the decision across contexts. (***DO1, DO2***) <br> Determine the timeframe within which the impact should be evaluated. (***DO1, DO2, DO4***) |

(*continued*)

Table 2. (continued)

| Evaluation concepts | Proposed evaluation requirements/considerations |
|---|---|
| Conformance | Determine the relevant conformance metrics and criteria across the contextual levels. (**DO1**)<br>Distinguish between short-term and long-term conformance evaluation criteria. Conformance requirements may change across time. (**DO1, DO2, DO4**) |
| Metrics | Deal with conflicting metrics, goals, and constraints. (**DO1**)<br>Incorporate separate metrics related to the human decision maker and the decision, along with the AI/ML metrics and those related to the machine. (**DO3**)<br>Prioritize the most important/relevant criteria and metrics and providing guidance on weights and selection of metrics. (**DO1, DO4**)<br>Differentiate between short-term and long-term evaluation metrics. Do not include all metrics each time. (**DO1, DO2, DO4**) |
| Errors and Biases | Differentiate between errors and biases related to human decision makers, machines (analytics), and data. (**DO1, DO3**)<br>Define appropriate metrics and criteria for evaluating errors and biases. (**DO1**)<br>Identify errors to learn from past decisions for future decisions. (**DO1, DO4**) |

## 4.2  Design Objectives for Ex-Post Decision Evaluation

Consequently, four main DOs were concluded for a future solution which responds to the needs of ChemML, as shown in Table 3.

The first DO for an implementable ex-post data-driven decision evaluation method is that it should be comprehensive and incorporate multi-faceted criteria. These criteria may range across the contextual levels priorly discussed, and include some of the proposed concepts as facets. For instance, in the ChemML case, the contextual levels can incorporate environmental impacts, which are governed by averaged or long-term process performance, whereas short-term decisions related to process operation may have positive short-term impacts (on a decision level) but negative long-term impacts. Furthermore, the criteria should differentiate between the data-driven decision elements, such as the evaluation of the machine, the decision outcome, the data, etc., which may potentially be conflicting and otherwise lead to confusion. In ChemML, although the accuracy and evaluation metrics of the ML tool's decision may have been high in a majority of instances, the expert's evaluation generally differed and took into account different aspects and criteria.

Similarly, DO2 encourages performing processual evaluation across different stages in time. This considers the changing contexts and aspects regarding the data-driven decision, which should be captured in the evaluation to understand the longitudinal consequences and impact of the decision. A concrete example related to ChemML would be the performance evaluation of indirect measurements, which can be dependent on factors such as seasonal variability of the raw materials, changes in ambient conditions, and unmodeled changes related to equipment fouling or degradation.

**Table 3.** Design objectives for ex-post evaluation of data-driven decisions

|   | Design objective |
|---|---|
| 1 | Incorporate multi-faceted (potentially conflicting) evaluation criteria across contextual levels |
| 2 | Perform processual evaluation across time |
| 3 | Define the applicable mode of collaboration between humans and machines and evaluate its effect on decision-making, decision outcomes, and collaborative rationality |
| 4 | Enable a (potentially automated) feedback loop for learning from the (discrete or continuous) evaluation of past decisions |

DO3 focuses on the relation between the human and the machine in the data-driven decision making process. By incorporating into the evaluation the mode of collaboration between humans and machines and the consequent effect on decision making, the decision outcomes, and achieving a collaborative rationality, we can glean more insights on such a collaboration and how to steer it to make better decisions. In ChemML, the evaluation would require, for example, regular interviews with end-users to assess the utilization degree of the machine, or development of automated logging of the human-machine interaction during the decision-making process. The latter could also facilitate DO4, where a (possibly automated) feedback loop ensues from the evaluation and enables learning through evaluating past decisions, both from an organizational and machine perspective, and consequently updates the training data to enhance ML. Similar to how decisions may be discrete or continuous, the evaluation of decisions and the resulting learning may also be discrete or continuous, depending on the decision type and context. Therefore, a design solution should be developed ingraining these objectives.

## 5   Discussion

The two main contributions of our research are:

1) the extraction of evaluation concepts from the literature, and
2) building upon them in the case of ChemML to define the requirements and DOs for data-driven decision evaluation in practice.

The concepts to be considered in ex-post decision evaluation are particularly of interest due to their ability to capture the multi-faceted and changing nature of data-driven decisions, rather than focus on individual or static evaluation concepts (e.g., at the level of the decision itself), as is mainly done in current studies. These concepts theoretically support, and are supported by, the practical example of ChemML. Its contemporary evaluation methods did not consider multiple criteria or contexts, although a comprehensive, ex-post evaluation method was desired to enable learning, as well as to enhance future decision making and increase adoption of the ML tool.

The DOs further contribute to theory and practice, and emphasize the need for a comprehensive evaluation method which incorporates multi-faceted evaluation criteria across the levels of the decision itself, organization, environment, and time. By reflecting on ChemML, we can see that multiple interrelated factors are present in each decision, and individual evaluation metrics on a single level remain insufficient in terms of ex-post learning. This challenges current research, which focuses on ML evaluation metrics, such as confidence, uncertainty, specificity, sensitivity, accuracy, area under the curve (AUC), etc. [26, 49]. Whereas such measures are necessary for evaluating the ML model performance as such, our paper argues that they are insufficient for ex-post evaluation and learning about the decisions.

This argument is in line with Lebovitz et al. [19], which show the limitations of primary performance measures used by managers to evaluate AI tools and their output. Contrarily, the actual results and knowledge of the experts, in many instances, conflict with the reported measures of the tools [16], which in the ChemML case decreased trust in the tool. Furthermore, the machine ignores certain important variables only human experts are capable of considering [14, 19], which was also the case with ChemML. This emphasizes the need for additionally accounting for the modes of collaboration between humans and machines in the evaluation of data-driven decisions.

Depending on the use case and level of analysis, one performance measure may be more important than another, and the mathematically optimal may become ethically problematic. Decision outcomes are thus the ultimate indicators of success and multiple factors should be considered in the evaluation, along with long-term follow up [19]. Accordingly, our first and third DOs support, and are supported by, such claims in recent research and endeavor to provide a solution to the evaluation paradox. Although some papers do consider evaluation of the algorithms, along with evaluation of the impact of the decision [14], their research focuses on a particular approach for data-driven decision making in a domain-specific decision, and they do not aim to provide ex-post evaluation solutions.

The second DO highlights the importance of a process science perspective and capturing the changes in contexts, concepts, and consequences, as well as understanding how they evolve, interact, and unfold, through a processual evaluation across time [5]. The set of decisions and concepts involved in the evaluation, as well as the evaluation method, may differ according to the stage in time when the evaluation is made. For example, in ChemML, production was evaluated within a shorter time frame based on whether the production targets were met. However, the environmental impact is used to evaluate a set of decisions at a later stage in time. Thus, it is crucial to know what to evaluate when.

The fourth DO builds on the traditional claim that ex-post evaluation enables learning from past decisions. This accentuates the need for designing a feedback loop which performs an evaluation based on the first two DOs, and feeds the results of the evaluation back into the process to enable a combination of both organizational and machine learning. This feedback loop is part of a prospective solution for monitoring how data-driven decisions are taken, cultivating criteria to evaluate such decisions, and reflecting through double-loop learning for the continuous evaluation and improvement of human-machine collaboration [7, 8, 20, 42]. While this feedback loop (or parts of it) could potentially be automated to simplify the task, we still support Grønsund and Aanestad's [12] claims that necessitate the human-in-the-loop configuration for ensuring that performance of the algorithm meets the organization's requirements.

Utilizing the knowledge presented by these DOs, a theory-ingrained and practically feasible solution to the ex-post evaluation of data-driven decisions can be developed. This further contributes to practice by enabling the evaluation and understanding of data-driven decisions, enhancing learning usage of AI and ML tools, and adding insights to the collaboration between humans and machines and the impact on decision making. Accordingly, decision makers, developers, and collaborators in the data-driven decision making process can benefit from the results. Finally, the development of a data-driven decision evaluation solution following the determined DOs may potentially address the data-driven decision making challenges faced by ChemML and many other organizations.

## 6  Conclusion and Future Work

In this paper, we aimed to problematize the ex-post evaluation of collaborative data-driven decisions, from the perspectives of theory and practice, and determine the DOs for a solution. Accordingly, by perusing the literature we determined the need for ex-post evaluation and a variety of concepts and factors to consider in the evaluation. From a practical perspective, ChemML exemplified the need for data-driven decision evaluation in industry, and was used to identify the necessary requirements and considerations for a proposed ex-post evaluation solution.

From these requirements, four DOs for a solution were proposed: (1) the existence of an implementable, comprehensive method incorporating multi-faceted (potentially conflicting) evaluation criteria across contextual levels (decision, organization, environment), (2) accounting for the changes in concepts, contexts, and outcomes across time and supporting a processual evaluation, (3) incorporating into the evaluation the mode of collaboration between humans and machines and its effect on decision-making, decision outcomes, and achieving a collaborative rationality, and (4) enabling a (potentially automated) feedback loop for learning from the evaluation of past decisions.

Future work includes utilizing the DOs towards building and testing a design artifact, in collaboration with industry, which could be used in an organizational context for the purpose of data-driven decision evaluation. This artifact should support "how" (process, metrics, and criteria) and "when" (which stages in time, if at all) to evaluate data-driven decisions. Additionally, we aim for a longitudinal case study in order to understand the organizational context surrounding data-driven decisions prior to the introduction of the evaluation, during the implementation, and post- implementation, following Bailey and Barley's [3] approach to studying intelligent systems in organizational contexts. Finally, we intend to research the concept of collaborative rationality further, and how to enhance the collaboration between humans and machines in decision making.

**Acknowledgment.** This research has been partially funded by the ITEA3 project Oxilate (https:// itea3.org/project/oxilate.html).

# References

1. Ajzen, I.: The social psychology of decision making. In: Social Psychology: handbook of basic principles, pp. 297–325 (1996)
2. Argyris, C., Schön, D.A.: Organizational Learning: A Theory of Action Perspective. Addison-Wesley. 77/78, 345 (1997). https://doi.org/10.2307/40183951
3. Bailey, D.E., Barley, S.R.: Beyond design and use: How scholars should study intelligent technologies. Inf. Organ. **30**(2), 100286 (2020). https://doi.org/10.1016/j.infoandorg.2019. 100286
4. Bouyssou, D. (ed.): Evaluation and Decision Models: A Critical Perspective. Kluwer Academic Publishers, Boston (2000)
5. vom Brocke, J., et al.: Process Science: The Interdisciplinary Study of Continuous Change. Social Science Research Network, Rochester, NY (2021). https://doi.org/10.2139/ssrn.391 6817
6. Chiang, L., et al.: Big data analytics in chemical engineering. Ann. Rev. Chem. Biomol. Eng. **8**(1), 63–85 (2017). https://doi.org/10.1146/annurev-chembioeng-060816-101555
7. Duan, Y., et al.: Artificial intelligence for decision making in the era of Big Data – evolution, challenges and research agenda. Int. J. Inf. Manage. **48**, 63–71 (2019). https://doi.org/10. 1016/j.ijinfomgt.2019.01.021
8. Dwivedi, Y.K., et al.: Artificial Intelligence (AI): multidisciplinary perspectives on emerging challenges, opportunities, and agenda for research, practice and policy. Int. J. Inf. Manage. **57**, 101994 (2021). https://doi.org/10.1016/j.ijinfomgt.2019.08.002
9. Elgendy, N., et al.: DECAS: a modern data-driven decision theory for big data and analytics. J. Decis. Syst. **31**, 1–37 (2021). https://doi.org/10.1080/12460125.2021.1894674
10. Gigerenzer, G., Gaissmaier, W.: Decision making: nonrational theories. In: International Encyclopedia of the Social & Behavioral Sciences, pp. 911–916. Elsevier (2015). https://doi.org/ 10.1016/B978-0-08-097086-8.26017-0
11. Gigerenzer, G., Gaissmaier, W.: Heuristic decision making. Ann. Rev. Psychol. **62**(1), 451–482 (2011). https://doi.org/10.1146/annurev-psych-120709-145346
12. Grønsund, T., Aanestad, M.: Augmenting the algorithm: Emerging human-in-the-loop work configurations. J. Strat. Inf. Syst. **29**(2), 101614 (2020). https://doi.org/10.1016/j.jsis.2020. 101614
13. Grover, V., et al.: The perils and promises of big data research in information systems. J. Assoc. Inf. Syst. **21**(2), 9 (2020). https://doi.org/10.17705/1jais.00601
14. Herm-Stapelberg, N., Rothlauf, F.: The crowd against the few: measuring the impact of expert recommendations. Decis. Support Syst. **138**, 113345 (2020). https://doi.org/10.1016/ j.dss.2020.113345
15. Hevner, A.R., et al.: Design science in information systems research. MIS Q. **28**(1), 75–105 (2004). https://doi.org/10.2307/25148625
16. Ioannidis, J.P.A., et al.: Forecasting for COVID-19 has failed. Int J Forecast. **38**, 423–438 (2020). https://doi.org/10.1016/j.ijforecast.2020.08.004
17. Klecun, E., Cornford, T.: A critical approach to evaluation. Eur. J. Inf. Syst. **14**(3), 229–243 (2005). https://doi.org/10.1057/palgrave.ejis.3000540
18. Kotsiantis, S.B., et al.: Machine learning: a review of classification and combining techniques. Artif. Intell. Rev. **26**(3), 159–190 (2007). https://doi.org/10.1007/s10462-007-9052-3

19. Lebovitz, S. et al.: Is AI ground truth really true? The dangers of training and evaluating AI tools based on experts' know-what. MIS Q. **45**(3), 1501–1526 (2021). https://doi.org/10.25300/MISQ/2021/16564
20. Lyytinen, K., et al.: Metahuman systems = humans + machines that learn. J. Inf. Technol. **36**(4), 427–445 (2020). https://doi.org/10.1177/0268396220915917
21. Magrabi, F., et al.: Artificial intelligence in clinical decision support: challenges for evaluating AI and practical implications: a position paper from the IMIA technology assessment & quality development in health informatics working group and the EFMI working group for assessment of health information systems. Yearb Med. Inform. **28**(01), 128–134 (2019). https://doi.org/10.1055/s-0039-1677903
22. March, J.G.: Bounded Rationality, Ambiguity, and the Engineering of Choice. The Bell Journal of Economics. **9**(2), 587–608 (1978). https://doi.org/10.2307/3003600
23. March, J.G.: Primer on Decision Making: How Decisions Happen. Simon and Schuster (1994)
24. Masha, E.M.: The case for data driven strategic decision making. Eur. J. Bus. Manage. **6**, 137–146 (2014)
25. Namvar, M., Intezari, A.: Wise data-driven decision-making. In: Dennehy, D., Griva, A., Pouloudi, N., Dwivedi, Y.K., Pappas, I., Mäntymäki, M. (eds.) Responsible AI and Analytics for an Ethical and Inclusive Digitized Society. LNCS, vol. 12896, pp. 109–119. Springer, Cham (2021). https://doi.org/10.1007/978-3-030-85447-8_10
26. Nasir, M., et al.: Developing a decision support system to detect material weaknesses in internal control. Decis. Support Syst. **151**, 113631 (2021). https://doi.org/10.1016/j.dss.2021.113631
27. Nutt, P.C., Wilson, D.C.: Handbook of Decision Making. John Wiley, Chichester (2010)
28. Peffers, K., et al.: A design science research methodology for information systems research. J. Manag. Inf. Syst. **24**(3), 45–77 (2007). https://doi.org/10.2753/MIS0742-1222240302
29. Peffers, K., Rothenberger, M., Kuechler, B. (eds.): Design Science Research in Information Systems. Advances in Theory and Practice. LNCS, vol. 7286. Springer, Heidelberg (2012). https://doi.org/10.1007/978-3-642-29863-9
30. Pettigrew, A.M.: Contextualist research and the study of organizational change processes. Res. Inf. Syst. **1**, 53–78 (1985)
31. Phillips-Wren, G., et al.: Cognitive bias, decision styles, and risk attitudes in decision making and DSS. J. Decis. Syst. **28**(2), 63–66 (2019). https://doi.org/10.1080/12460125.2019.1646509
32. Power, D.J., et al.: Analytics, bias, and evidence: the quest for rational decision making. J. Decis. Syst. **28**(2), 120–137 (2019). https://doi.org/10.1080/12460125.2019.1623534
33. Ransbotham, S. et al.: Expanding AI's Impact With Organizational Learning. https://sloanreview.mit.edu/projects/expanding-ais-impact-with-organizational-learning/. Accessed 22 Dec 2021
34. Raschka, S.: Model Evaluation, Model Selection, and Algorithm Selection in Machine Learning. http://arxiv.org/abs/1811.12808 (2020)
35. Russell, S., Norvig, P.: Artificial Intelligence: A Modern Approach. Pearson, Hoboken (2021)
36. Shrestha, Y.R., et al.: Organizational decision-making structures in the age of artificial intelligence. Calif. Manage. Rev. **61**(4), 66–83 (2019). https://doi.org/10.1177/0008125619862257
37. Simon, H.A.: A behavioral model of rational choice. Q. J. Econ. **69**(1), 99 (1955). https://doi.org/10.2307/1884852
38. Smith, J.A.: Qualitative Psychology: A Practical Guide to Research Methods. SAGE, Thousand Oaks (2015)
39. Snowden, D.J., Boone, M.E.: A leader's framework for decision making. Harv. Bus. Rev. **85**, 68 (2007)

40. Sonnenberg, C., vom Brocke, J.: Evaluations in the science of the artificial – reconsidering the build-evaluate pattern in design science research. In: Peffers, K., Rothenberger, M., Kuechler, B. (eds.) Design Science Research in Information Systems. Advances in Theory and Practice. LNCS, vol. 7286, pp. 381–397. Springer, Heidelberg (2012). https://doi.org/10.1007/978-3-642-29863-9_28

41. Stockdale, R., Standing, C.: An interpretive approach to evaluating information systems: a content, context, process framework. Eur. J. Oper. Res. **173**(3), 1090–1102 (2006). https://doi.org/10.1016/j.ejor.2005.07.006

42. Sturm, T., et al.: Coordinating human and machine learning for effective organization learning. MISQ. **45**(3), 1581–1602 (2021). https://doi.org/10.25300/MISQ/2021/16543

43. Sturm, T., et al.: The Case of Human-Machine Trading as Bilateral Organizational Learning. 18 (2021)

44. Tomperi, J., et al.: Mass-balance based soft sensor for monitoring ash content at two-ply paperboard manufacturing. Nord. Pulp Pap. Res. J. **37**(1), 175–183 (2022). https://doi.org/10.1515/npprj-2021-0046

45. Troisi, O., et al.: Growth hacking: Insights on data-driven decision-making from three firms. Ind. Mark. Manage. **90**, 538–557 (2020). https://doi.org/10.1016/j.indmarman.2019.08.005

46. Trunk, A., Birkel, H., Hartmann, E.: On the current state of combining human and artificial intelligence for strategic organizational decision making. Bus. Res. **13**(3), 875–919 (2020). https://doi.org/10.1007/s40685-020-00133-x

47. Tversky, A., Kahneman, D.: Rational choice and the framing of decisions. J. Business. **59**(4), S251–S278 (1986)

48. Tversky, A., Kahneman, D.: The framing of decisions and the psychology of choice. Science **211**(4481), 453–458 (1981). https://doi.org/10.1126/science.7455683

49. Vo, N.N.Y., et al.: Deep learning for decision making and the optimization of socially responsible investments and portfolio. Decis. Support Syst. **124**, 113097 (2019). https://doi.org/10.1016/j.dss.2019.113097

50. van Voorst, S., Zwaan, P.: The (non-)use of ex post legislative evaluations by the European commission. J. Eur. Publ. Policy **26**(3), 366–385 (2019). https://doi.org/10.1080/13501763.2018.1449235

51. Weick, K.E.: Sensemaking in Organizations. Sage Publications, Thousand Oaks (1995)

52. Weirich, P.: Realistic Decision Theory: Rules for Nonideal Agents in Nonideal Circumstances. Oxford University Press, Oxford (2004)

# Designing XML Schema Inference Algorithm for Intra-enterprise Use

Dmitry Uraev$^{(\boxtimes)}$ and Eduard Babkin

HSE University, Nizhny Novgorod, Russia
{duraev,eababkin}@hse.ru

**Abstract.** The paper considers methods for automated XML schema inference for a collection of documents from the point of the integration and usage at the enterprise. Most existing algorithms work with certain XML data that is claimed to be without errors or inaccuracies in the collection. The paper analyzes the theoretical foundations for inferring XML schema for decision maker who is working with automatically or manually created heterogeneous data. An algorithm based on a probabilistic approach is supposed to work on any data and allows the decision maker to have alternatives with a certain confidence level when working with an XML schema inferred. As a result of our findings, we introduce *xml.schema.inference* application for inferring XML schemas for intra-enterprise use.

**Keywords:** Probabilistic XML · Schema inference · Heterogeneous data · Uncertain data · xml.schema.inference · Schema editing

## 1 Introduction

The task of effectively supporting documentation at an enterprise, whether it is internal document management or the technical documentation created for a product, always plays an important role in any organization. Every day, the amount of data is steadily growing, and this data is becoming more and more difficult to process manually, as well as by automated means. Moreover, with the growth of the amount of data, the complexity of transferring this data between various subsystems at the enterprise also grows [7]. One of the most common formats for storing a collection of documents at the enterprise is the XML markup language, which is a format that provides a common interface for exchanging data between different subsystems.

Extensible markup language (XML) developed and maintained by W3C is an accepted standard for representing continuously increasing data both in the Internet environment in general and at the enterprise in particular [23]. The markup language allows not only storing huge amounts of data, but also managing it efficiently, e.g. by integrating data from several sources into one. In addition to storing structured data and being able to exchange information between programs, XML can also be used to create specialized derivative languages based

© The Author(s), under exclusive license to Springer Nature Switzerland AG 2022
E. Nazaruka et al. (Eds.): BIR 2022, LNBIP 462, pp. 35–49, 2022.
https://doi.org/10.1007/978-3-031-16947-2_3

on it. There are also EDIFACT or ANSI ASC standardization approaches that can be used to solve similar problems. However, XML is one of the newest and one of the most popular solutions, it can deliver more opportunities by flexibility, greater user-friendliness, and the ability to work in different languages' (Unicode) features.

Researchers repeatedly consider the possibility of introducing XML technologies to expand the functionality of the document management system at an enterprise [6,7]. However, when introducing such technologies at the enterprise, a number of problems arise that need to be resolved for the correct functioning of the informational systems as well as for the decision making purposes. First of all, these are the issues of integrating XML data from various sources. Researchers have proposed many algorithms and approaches for accurate and high-quality merging and inferring of XML data from different domains or systems [3–5,8,10,12,15,18,19,21,22]. However, the problem is that such algorithms perform effectively on certain data, which significantly reduces the range of possibilities for applying such algorithms at the enterprise. At the same time, studies [3] show that approximately only half of the XML documents on the Web refer to a schema corresponding to them. Additionally, the research in [20] shows that third of the schemas on the Web are ambiguous that is semantically or grammatically incorrect. Most XML algorithms and approaches are focused on working with certain data (deterministic XML), in which the structure in various documents and data sources is clear and does not imply any inaccuracies, errors, or deviations from the general structure [4]. However, data is inherently ambiguous ('fuzzy', 'uncertain'), due to which, when integrating data or solving other problems, conflicts or errors in the operation of algorithms may occur (e.g. errors due to the human factor and/or errors due to describing the entity in documents in different words), which also makes it difficult to use such solutions at the enterprise. Such issues are considered by researchers in works on ambiguous XML data (non-deterministic XML approach) [1,2,10–12,15,22].

The use of the XML standard at an enterprise is associated with the constant need to process a large amount of data, either collected by automated systems or created and edited by a human [7]. The primary need for a decision maker who works with collections of such data is to be able to have confidence in the data they receive from the storage. One way is to look at the XML schema that describes this particular XML collection. However, most algorithms set as the primary task the output of an unambiguous schema or several variants of the schema leaving all situations of decision-making about the entry of one or another element into the schema within the algorithm or as a part of user interaction during the before-inferring step [1,2,10–12,15,22]. In such a case, the decision maker receiving the schema must rely on the correct operation of the schema inference algorithm or the person who interacted with schema inference algorithm during editing step. However, what if we can give the decision maker the opportunity to look at the schema, which presents a set of alternatives, from which it is possible, with varying degrees of confidence, to select the correct elements to use in the resulting schema.

Another problem with using the XML standard at an enterprise, besides the difficulty of a decision maker working with a schema representing the data, is also the strong domain specificity of the enterprise tasks for which the schema inference algorithm needs to be used [7]. There may be a question of trust in the work of one or another algorithm for inferring a schema from a collection of data in the case of an inference of heterogeneous data. Moreover, inference algorithms need to be specialized and redesigned for specific tasks, which raises the cost of potentially reusing such algorithms in other parts of the enterprise.

Hence, the main objective of the current study is to elaborate a simple XML schema inference algorithm and practical implementation of it for a decision maker at an enterprise in order to provide the ability to select the structural elements of the XML schema from a variety of alternatives at the stage of working with the inferred XML schema, and also to eliminate the need to highly specialize the schema inference algorithm for a specific task, which shall lead to increase in re-usability of the algorithm in different parts of the enterprise. We also make an attempt to broaden existing method for the probabilistic inference of XML schema and to transfer existing approach to a related problem of working with XML, that is from probabilistic XML integration to probabilistic XML schema inference.

## 2   Literature Overview

After defining the problem, we focused on the analysis of existing approaches for the automatic XML schema inference, as well as on direct analysis of the existing developed solutions.

### 2.1   Structural-Semantic XML Approach

The most common approach when working with XML data consists of two main parts - areas of development when working with XML: the structure of the data and the content of the data. In other words, within the framework of the approach for parsing XML documents and, in particular, deriving an XML schema from a collection of documents, the location of tags in the hierarchy of each individual document is analyzed. The positions of the tags are analyzed and compared with each other using further manipulations to derive knowledge about the location of these tags in the data [15]. For example, the decision to choose one or another tag in the data hierarchy can be made using a decision algorithm based on regular expressions [9] where the characters *, ?, +, etc. have the same semantics as in the usual regular expressions, however, not letters but whole tags are used as the units under consideration.

The semantic part of the approach usually consists of developing algorithms for comparing each pairs of two tags in a collection. The result of such a comparison is knowledge about the correspondence of two tags to each other or their complete mismatch. At this stage, various auxiliary algorithms can be used, ranging from the simplest, such as edit distance, to complex algorithms using word embedding. More details about the methods of semantic analysis of the elements of a collection of documents are described in [18]. The complexity and expediency of applying certain similarity algorithms in different situations is described in [9].

## 2.2  Probabilistic XML Approach

The probabilistic approach for storing and processing XML data is based on the concept of uncertain or fuzzy data. Such data can be obtained both through automated data collection tools, e.g. in the Web, and manual means, e.g. data on patient diagnoses in a clinic. The probabilistic approach allows user to determine which data element is a fact and which element arose as a result of the inaccuracy of the data collection algorithm or as a result of a human factor with some probability. Moreover, the probabilistic approach offers a generalized model for storing and manipulating ambiguous data that makes it possible to work effectively with this data not only within specific-domain tasks and highly specialized programs but also other within tasks from different areas of application [12].

Among the main areas in which the probabilistic approach is used for storing and processing ambiguous data, researchers identify: merging XML data from various sources, extracting information from the Web, parsing or building syntax trees, collaborative editing of common documents, data queries to large corpora [12]. All of these tasks involve dealing with ambiguous data to a greater or lesser extent. For example, when editing the same document by multiple authors, a single edit operation by the author may be considered an ambiguous operation. This raises the question of how much this operation can be trusted and whether this operation violates the rules for co-writing a document for all authors. It is worth noting that the task of deriving an XML schema from a given collection of XML documents is also included in the list of tasks that can be solved by using a probabilistic approach, since each individual document in the collection can be considered as a separate probability distribution. To do so, we need to build a general probability distribution over the collection complementing to the fact that any document may have ambiguous data. However, this problem is still open for solution, since it has both theoretical and implementation difficulties, as researchers noted in [12].

## 2.3  Heuristic XML Approach

A group of methods that is also worth noting and is not included in the previously identified approaches is a group of heuristic methods. These approaches are opposed to other methods that consider XML documents as trees with a certain hierarchical structure and labels, e.g. grammar-based view. As the researchers note in [16,17], heuristic approaches form a separate group of methods, since it is impossible to give unambiguous characteristics and descriptions of the patterns of their work from the point of view of the theory of markup languages. However, such approaches are still used in practical application since they have intuitively understandable and natural inferring algorithms as stated in [17].

The general task of deriving an XML schema using heuristic approaches still comes down to finding such a schema that will be short enough and accurately describing a given collection of documents while having a good ability to generalize that is not being too specific to a particular collection of texts [4,5,10]. In this case, several potential schemas may exist at once, so the problem is also expanded by searching for the optimal scheme among all options [17].

### 2.4 Literature Review Findings

Based on the results of the analysis of existing approaches, several remarks can be made. First of all, most of the approaches, although singled out and described separately, are also implemented in practice in conjunction with other approaches. Thus, when working with a XML collection of documents and analyzing the structure and content of the collection according to the structural-semantic approach, decision algorithms can still resort to heuristic indicators to make decisions about the inclusion of one or another element in the final schema. As it was written earlier, it is heuristic approaches that are most popular in the XML schema output task which have implementation steps that are intuitively understandable to the user but do not have high theoretical reliability. As for the structural-semantic approach, the method also finds its application in the problem of schema derivation, however, due to its specificity and strong dependence on the nature of the data, it shows good results mostly on certain data and requires high domain specification for uncertain data inference. The probabilistic approach, although comprehensively described by the theoretical framework, is used only in a number of tasks related to XML and was not sufficiently covered and transferred to the task of inferring a schema from a collection of documents. Moreover, existing probabilistic approaches offer potential XML schemas based on the results of their work without allowing the decision-maker to edit the final schema. Algorithms are focused on deriving a generalized and at the same time accurate schema for specific data that may lead to an unsuccessful user experience for a decision maker at an enterprise when working with the resulting schemas in the case of searching for alternatives, when working with uncertain data.

When developing a solution for the current work, we will primarily focus on works on the probabilistic inference of the XML schema since it is with the help of probabilities that, in our opinion, it is possible to implement a set of alternatives for the decision maker who is working with the inferred schema. Key works on probabilistic XML inference that we will mainly focus on and refer to are [1,2,10–12,15,22]. We will also use other approaches solutions if needed with the corresponding explanation.

## 3   The Approach Proposed

Let us consider a collection of XML documents $D$ that consists of tags $t_i$, $i = 1...n$, $n \in N$, attributes $a_j$, $j = 1...n$, $n \in N$, and their content marked as constant $C$. The collection can be considered as the finite set of documents $d_1, d_2, ...d_n \in D$. By iterating over each document $d_k$ in the collection, we can apply count function to get all occurrences of each tag $t_i$, attribute $a_j$ in the document. Tag occurrence $O$ of tag $t_i$ is formalized as follows $O_{t_i} = count(d_k, t_i), k = 1...n$, where $n$ is the last document in the collection, $i = 1...m$, where $m$ is the last tag in particular document $d_k$. The occurrence $O$ of attribute $a_j$ is formalized as follows $O_{a_j} = count(d_k, a_j), k = 1...n$, where $n$ is the last document in the collection, $j = 1...m$, where $m$ is the last attribute in particular document $d_k$. The overall element $l$ (tag or attribute) occurrence in the collection is the sum of occurrences of this element $l$ in each document of the collection:

$$O_l = \sum(O_1, ...O_n) \tag{1}$$

where $n$ is the occurrence of the element $l$ in the last document in the collection $D$. Hence, we can go further and count the probability $p$ of the element $l_i$ in a specific document $d_j$:

$$p(d_j, l_i) = \frac{count(d_j, l_i)}{O_{l_i}} \tag{2}$$

Thus, each tag and attribute in each document of the collection is assigned with the probability $p$. This probability is used in the XML schema inference algorithm for inference purposes as well as translated into the final XML schema for decision maker evaluation in the form of sets of alternatives.

The overall inference algorithm can be described with the following stages:

(1) Derivation of *initial grammar* that is getting all the tags, attributes, and content from the collection.
(2) Counting the occurrences and probabilities of each element in the collection space and assigning this information to the corresponding elements.
(3) Building *production rules* using *initial grammar* and probabilities.
(4) Grouping and merging *production rules* according to probabilities with the help of heuristics.
(5) Collecting data types and additional information for the schema. Defining sets of alternatives based on probabilities for decision maker.
(6) Obtaining the complete internal representation of collection in the form of production rules.
(7) Inferring XML schema from the internal representation including sets of alternatives for the decision maker.

Figure 1 shows the whole process of inferring XML schema from the collection of XML documents. The overall algorithm complements to the existing methods described in [13,15–17] using the same terminology, modelling, and internal data representation concepts. On the other hand, we modify general approach, including probabilities and sets of alternatives processing during the algorithm run.

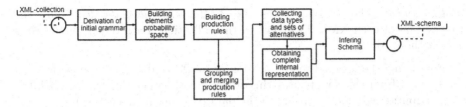

**Fig. 1.** The process of inferring XML schema from collection of XML documents using elements probabilities and occurrences.

## 4    Architecture of the Application Developed

To support our approach proposed, we implemented the *xml.schema.inference* application. The application is written using Python programming language and is distributed via GitHub. *xml.schema.inference* consists of four main modules and two additional utility modules, 730 lines of code in total. The Fig. 2 shows the overall architecture of the application developed.

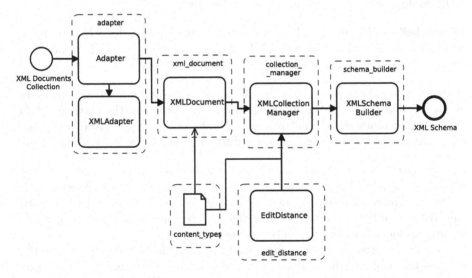

**Fig. 2.** *xml.schema.inference* application architecture.

*Adapter.* The first block in Fig. 2 is presented as a standalone module that operates on an incoming collection of XML documents. At the moment, the interface allows you to open as separate XML files or open the entire collection using the path to the folder with files. Each XML document is opened and converted to an internal syntax-tree representation using the built-in standard *xml* library.

*xml_document.* The module stores the *XMLDocument* class describing the abstraction of each XML document in the collection. Each object of the class has methods for:

1. Populating the internal representation of the XML document using the data received from the adapter (at this stage, among other things, a probability distribution is built according to the data of a specific document).
2. Getting the initial grammar for a specific document.
3. Simplifying initial grammar within a particular document (clustering and removing duplication).

*collection_manager.* *XMLDocument* objects are created for each document in the collection in the module using the *XMLCollectionManager* class. The module works with the entire collection and is used for:

1. Calculation of probability distributions for each element in the collection.
2. Updating the relative frequencies for the collection.
3. Assembling of production rules.
4. Processing all production rules for the collection - grouping and removing duplicates, analyzing sets of alternatives.
5. Building summarized internal data representation.

*schema_builder.* All received data as a result of the work of the collection manager is sent to the module in with *XMLSchemaBuilder* class which is responsible for creating a schema file in the current working directory.

*content_types.* The module is an auxiliary file that stores the necessary data about existing attribute types for algorithms to decide on the attribute data type. At the moment, a small number of data types are provided but other types are planned to be added as part of further improvements of the algorithm.

*edit_distance.* The additional module with the implemented *EditDistance* class is used by the main algorithm at the stages of analysis of alternative sets for the collection. The module implements the work of the algorithm for calculating the edit distance of two strings-tags or attributes. The module is used among other algorithms for finding multiple alternatives.

To run the program, user need to load the collection of documents into the samples folder or call the appropriate adapter module that opens the collection from any path on the user's computer. Next, to process the collection and display the schema, user needs to run the $collection_{m}anager$ module. The final file with the scheme will be in the program launch directory.

Beyond the scope of the current description, there are also private methods that are directly involved in calculating key parameters for a collection analyzing tags and attributes and building sets of alternatives for decision maker. We plan to give a detailed description of the methods for presenting data within the program, as well as the decision algorithms themselves, in our next work. Most of the methods are written from scratch and are in the file with the modules themselves.

# 5  Evaluation and Validation

After the implementation of the main program code, the operation of the algorithm was tested on a small test data set prepared (see Fig. 3). As a practical application case, the situation of collecting reports on a network of general-purpose stores was chosen. The situation is simulated when cashiers enter data on purchased goods during the working day. The report provides a snapshot of the situation at a specific point in the network for a given period of time. Depending on the systems used to store such data on sales of goods, users may have different levels of access to the original data which can be stored including in XML format.

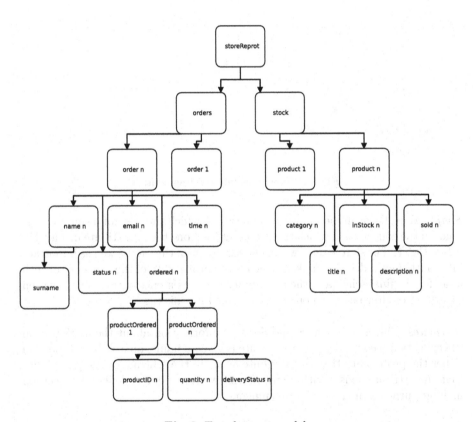

**Fig. 3.** Test data set model.

As a starting point for generating a data set, we created an XSD schema that represents all the features of the subject area. 10 XML documents were generated according to the given schema using generating algorithms. The solution provided in the trial version of the Liquid Studio XML editor was used as a generating algorithm [24]. During the generation of test XML documents, some of them were intentionally generated with incorrect and erroneous data to test the algorithm (we intentionally corrupted schema by changing data structure, corrupting tag names etc.). Errors in the data were of two types: 1) errors in the name of a tag or attribute that are errors in content, and 2) errors affecting the structure of documents, making it incorrect. In the following paragraphs, we briefly describe the schema derived for the XML test data set pointing out the key features including the display of alternative sets.

```
<!-- Simple Elements -->
<!-- probability of surname present is 0.5, otherwise it is not used -->
<xs:element name="surname" type="xs:string"/>
<!-- probability of email present is 1.0, otherwise it is not used -->
<xs:element name="email" type="xs:string"/>
<!-- probability of quantity is 0.9565217391304348, other options: [('quantiti', 0.043478260869565216)] -->
<xs:element name="quantity" type="xs:string"/>
<!-- probability of time present is 0.8, otherwise it is not used -->
<xs:element name="time" type="xs:string"/>
<!-- probability of inStock present is 1.0, otherwise it is not used -->
<xs:element name="inStock" type="xs:string"/>
<!-- probability of sold present is 1.0, otherwise it is not used -->
<xs:element name="sold" type="xs:string"/>
<!-- probability of title present is 1.0, otherwise it is not used -->
<xs:element name="title" type="xs:string"/>
<!-- probability of description present is 1.0, otherwise it is not used -->
<xs:element name="description" type="xs:string"/>
<!-- probability of quantiti is 0.043478260869565216, other options: [('quantity', 0.9565217391304348)] -->
<xs:element name="quantiti" type="xs:string"/>
```

**Fig. 4.** Simple tags inferred fragment.

*Simple Tags.* By a simple tag we mean a tag for which there is not a single occurrence in the collection in which the tag has at least one tag or attribute included in it (see Fig. 4). For simple tags, we display the name of the tag, as well as important information for a decision maker, which is the probability of the existence of such a tag for the final schema or the probability of the existence of this tag in relation to any set of alternatives, if one can be defined for this tag.

*Attributes.* This group defines and displays information about the attributes in the collection (see Fig. 5). Each attribute is given a name, an attribute type, and either the probability that the attribute exists in the schema, or the probability that the attribute exists within any alternative set, if one is defined for decision making purposes in the form of comments.

*Semi-simple Elements.* This group includes those tags that do not contain other included tags but contain one or more attributes (see Fig. 6). For semi-simple elements, information is displayed about the included attributes (in the form of links to the previously indicated attributes of the corresponding group), as well as the probability of using semi-simple tags in the schema, or the probability of the existence of semi-simple elements within certain sets of alternatives, if any. Also, for each attribute within the framework of semi-simple tag, information is

```
<!-- Attributes -->
<!-- probability of order_id present is 1.0, otherwise it is not used -->
<xs:attribute name="order_id" type="xs:nonNegativeInteger"/>
<!-- probability of order_status present is 1.0, otherwise it is not used -->
<xs:attribute name="order_status" type="xs:string"/>
<!-- probability of product_id present is 0.8, otherwise it is not used -->
<xs:attribute name="product_id" type="xs:nonNegativeInteger"/>
<!-- probability of product_status present is 0.8, otherwise it is not used -->
<xs:attribute name="product_status" type="xs:string"/>
<!-- probability of product_stock_id present is 1.0, otherwise it is not used -->
<xs:attribute name="product_stock_id" type="xs:nonNegativeInteger"/>
<!-- probability of product_category present is 1.0, otherwise it is not used -->
<xs:attribute name="product_category" type="xs:string"/>
<!-- probability of product_additional_category is 0.9090909090909091,
other options: [('product_aditional_category', 0.09090909090909091)] -->
<xs:attribute name="product_additional_category" type="xs:string"/>
<!-- probability of product_aditional_category is 0.09090909090909091,
other options: [('product_additional_category', 0.9090909090909091)] -->
<xs:attribute name="product_aditional_category" type="xs:string"/>
```

**Fig. 5.** Attributes inferred fragment.

```
<!-- Semi-simple Elements -->
<!-- probability of status present is 1.0, otherwise it is not used -->
<xs:element name="status">
<xs:complexType>
<xs:attribute ref="order_status" use="required"/>
</xs:complexType>
</xs:element>
<!-- probability of productId present is 0.8, otherwise it is not used -->
<xs:element name="productId">
<xs:complexType>
<xs:attribute ref="product_id" use="optional"/>
</xs:complexType>
</xs:element>
<!-- probability of deliveryStatus present is 0.8, otherwise it is not used -->
<xs:element name="deliveryStatus">
<xs:complexType>
<xs:attribute ref="product_status" use="optional"/>
</xs:complexType>
</xs:element>
<!-- probability of category is 0.8928571428571429,
other options: [('categori', 0.10714285714285714)] -->
<xs:element name="category">
<xs:complexType>
<xs:attribute ref="product_category" use="required"/>
<xs:attribute ref="product_additional_category" use="optional"/>
<xs:attribute ref="product_aditional_category" use="optional"/>
</xs:complexType>
</xs:element>
```

**Fig. 6.** Semi-simple elements inferred fragment.

displayed about the need to use this attribute in a given position of the semi-simple tag. An attribute can be either mandatory (at least one occurrence) or optional.

```
<!-- Complex Elements -->
<!-- probability of name is 0.9333333333333333,
other options: [('names', 0.06666666666666667)] -->
<xs:element name="name">
<xs:complexType mixed="true">
<xs:all>
<xs:element ref="surname" minOccurs="0" maxOccurs="1"/>
</xs:all>
</xs:complexType>
</xs:element>
<!-- probability of product is 0.75, other options:
[('praduct', 0.21428571428571427), ('preduct', 0.03571428571428571)] -->
<xs:element name="product">
<xs:complexType mixed="true">
<xs:all>
<xs:element ref="category" minOccurs="0" maxOccurs="1"/>
<xs:element ref="inStock" minOccurs="1" maxOccurs="1"/>
<xs:element ref="sold" minOccurs="1" maxOccurs="1"/>
<xs:element ref="title" minOccurs="1" maxOccurs="1"/>
<xs:element ref="description" minOccurs="0" maxOccurs="1"/>
<xs:element ref="categori" minOccurs="0" maxOccurs="1"/>
</xs:all>
<xs:attribute ref="product_stock_id" use="required"/>
</xs:complexType>
</xs:element>
```

**Fig. 7.** Complex tags inferred fragment.

*Complex Tags.* This group includes tags that have included tags and may or may not have included attributes (see Fig. 7). The algorithm displays information about each complex tag, its content (links to the corresponding included elements), as well as information about the intervals of occurrence of included elements (*minOccurs, maxOccurs* for tags and optional or required for attributes). Also, for a number of complex tags, a certain sequence of included elements can be displayed, if such was found by the algorithm. In addition, for each complex tag, the probability of its existence in the final schema, or the probability of its existence within a certain set of alternatives, if such is determined by the algorithm, is displayed.

We checked the ability of algorithm to form sets of alternatives including both correct and uncertain data for decision maker. The algorithm showed 100% coverage for correct attributes and tags when displaying the elements of the test set. The final schema included both correct elements and elements that imply uncertainty. Only 63% of the incorrect elements (leading to ambiguity) were inferred. At the same time, 100% of the derived incorrect elements were a part of the proposed sets of alternatives which included the correct options of the elements. The algorithm showed good results on the test sample not allowing the loss of any of the necessary elements and, at the same time, not allowing the use of incorrect elements in the resulting schema due to low probability. In the future, we plan to test the algorithm on other data sets, including larger ones, as well as verify the results of work with other solutions.

# 6   Discussion

The algorithm we developed has no exact analogues in terms of schema output using probabilities at the core of the algorithm as well as the opportunities provided to the decision maker when working with the schema (many alternatives for choosing the correct options, the level of confidence for all elements of the output XML schema). The other solutions under consideration [5,13,14,19] do not offer the decision maker such schema editing capabilities. At the same time, the resulting algorithm follows the general principles from the list of generally accepted approaches [4,12,13] that use the terms of generative grammar and production rules as the basis for modeling. The developed algorithm works with the structure and content of XML data like other solutions [15] being also able to output any information contained in XML data. On the other hand, the developed algorithm does not yet support schema versioning as well as working with other data formats features that are implemented in other solutions [5,13,14,19] and can be added as part of further improvements. The closest analogue is the solution proposed in [5] that offers a schema output using the traditional approach in conjunction with regular expressions as well as a built-in module for viewing documents in collection which allows user to look at the probabilities of occurrence of elements in a collection and determine ambiguous elements in a collection, but this module only offers additional functionality that does not affect the output of a schema by a collection and is not reflected in the schema resulted. The key difference of our application is the use of a probabilistic approach at the core of the schema inference. The application is completely focused on the probabilities of elements when deriving a schema with the sufficient information for decision maker additionally resorting to heuristic indicators when building sets of alternatives.

With regard to Enterprise Application Integration (EAI), the developed solution broadens the decision maker's abilities. For example, schema inference can be built into the process of validating configuration documents for running servers in an enterprise. In such a case, the output schema from the collected configuration documents can be evaluated by the decision maker, and checked for compliance, or used as the basis for generating configurations on newly added servers. By viewing the same schema the decision maker is able to validate that information is configured and stored in the same conceptual way on different servers. If any low probability ejections happen, it is possible for the decision maker to validate and to solve the case. Another useful point of application for the developed solution is being able to convert data from different sources (e.g. Markdown, rST, HTML etc.) into XML using the schema provided. In this case, the decision maker again is able to get the current XML documents collection schema representing data structure and, after editing, to use the schema alongside the format converter.

# 7 Conclusion

In the current work we proposed an approach based on probabilities for inferring XML schema with possibility to show sets of alternatives for decision maker at the enterprise. To test and to support our approach, we developed *xml.schema.inference* application that allows user to process a collection of XML documents and display a schema with the possibility of seeing alternatives for decision making. We managed to implement the output of tags, attributes as well as additional information related to them: information about the type of data contained in the attribute, information about the structure of the tags. We plan to continue improving the developed algorithm as it has potential for improvements such as inclusion of new decision algorithms, creation of user interface for decision maker usage and a more intuitive schema output format. The work of the resulting algorithm also needs to be tested on other data sets including larger ones to identify possible cases of unstable or incorrect behavior.

# 8 xml.schema.inference Source Code

The source code of our application is privately available on GitHub and can be provided upon request by mail for research or development purposes.

# References

1. Abiteboul, S., Amsterdamer, Y., Deutch, D., Milo, T., Senellart, P.: Optimal probabilistic generators for xml corpora. In: BDA (Bases de données avancées), p. 20 (2011)
2. Benedikt, M., et al.: Probabilistic XML via Markov chains. Proc. VLDB Endow. **3**(1) (2010)
3. Barbosa, D., Mignet, L., Veltri, P.: Studying the XML web: gathering statistics from an XML sample. World Wide Web **8**(4), 413–438 (2005)
4. Bex, G.J., Neven, F., Vansummeren, S.: Inferring XML schema definitions from XML data. In: Proceedings of the 33rd International Conference on Very Large Data Bases, pp. 998–1009 (2007)
5. Bex, G.J., Neven, F., Vansummeren, S.: SchemaScope: a system for inferring and cleaning XML schemas. In: Proceedings of the 2008 ACM SIGMOD International Conference on Management of Data, pp. 1259–1262 (2008)
6. Besedina, K.V.: Osobennosti yazyka razmetki xml. Eur. Res. **8**(19), 51–52 (2016)
7. Borisenko, E.V.: XML-bazy dannykh v sisteme upravleniya elektronnym dokumentooborotom. Nauka i sovremennost, no. 4-2, pp. 248–253 (2010)
8. Gómez, S.N. et al.: Findings from Two Decades of Research on Schema Discovery using a Systematic Literature Review. AMW (2018)
9. Groz, B., et al.: Inference of Shape Expression Schemas Typed RDF Graphs. arXiv preprint arXiv:2107.04891 (2021)
10. Janga, P., Davis, K.C.: Schema extraction and integration of heterogeneous XML document collections. In: Cuzzocrea, A., Maabout, S. (eds.) MEDI 2013. LNCS, vol. 8216, pp. 176–187. Springer, Heidelberg (2013). https://doi.org/10.1007/978-3-642-41366-7_15

11. Kharlamov, E., Senellart, P.: Modeling, querying, and mining uncertain XML data. In: XML Data Mining: Models, Methods, and Applications, pp. 29–52. IGI Global (2012)
12. Kimelfeld, B., Senellart, P.: Probabilistic XML: models and complexity. In: Advances in Probabilistic Databases for Uncertain Information Management, pp. 39–66 (2013)
13. Klempa, M., et al.: JInfer: a framework for XML schema inference. Comput. J. **58**(1), 134–156 (2015)
14. Li, Y., et al.: FlashSchema: achieving high quality XML schemas with powerful inference algorithms and large-scale schema data. In: 2020 IEEE 36th International Conference on Data Engineering (ICDE), pp. 1962–1965 (2020)
15. Ma, Z., Zhao, Z., Yan, L.: Heterogeneous fuzzy XML data integration based on structural and semantic similarities. Fuzzy Sets Syst. **351**, 64–89 (2018)
16. Mlýnková, I.: XML Schema Inference: A Study. Technická zpráva, Charles University, Prague, Czech Republic (2008)
17. Mlýnková, I., Nečaský, M.: Heuristic methods for inference of XML schemas: lessons learned and open issues. Informatica **24**(4), 577–602 (2013)
18. Oliveira, A., et al.: An efficient similarity-based approach for comparing XML documents. Inf. Syst. **78**, 40–57 (2018)
19. Oliveira, A., et al.: XChange: a semantic diff approach for XML documents. Inf. Syst. **94**, 101610 (2020)
20. Sahuguet, A.: Everything you ever wanted to know about DTDs, but were afraid to ask (extended abstract). In: Goos, G., Hartmanis, J., van Leeuwen, J., Suciu, D., Vossen, G. (eds.) WebDB 2000. LNCS, vol. 1997, pp. 171–183. Springer, Heidelberg (2001). https://doi.org/10.1007/3-540-45271-0_11
21. Siméon, J., Wadler, P.: The essence of XML. ACM SIGPLAN Not. **38**(1), 1–13 (2003)
22. Van Keulen, M., De Keijzer, A., Alink, W.: A probabilistic XML approach to data integration. In: 21st International Conference on Data Engineering (ICDE 2005), pp. 459–470 (2005)
23. World Wide Web Consortium. XML Technology. http://www.w3.org/standards/xml/. Accessed 7 Jan 2022
24. Liquid Technologies. XML Liquid Studio. http://www.liquid-technologies.com/. Accessed 18 May 2022

# Modeling Methods and Assistance

# Endurant Ecosystems: Model-Based Assessment of Resilience of Digital Business Ecosystems

Jānis Grabis[1]([⊠]) [iD], Chen Hsi Tsai[2] [iD], Jelena Zdravkovic[2] [iD], and Janis Stirna[2] [iD]

[1] Information Technology, Riga Technical University, Kalku 1, Riga 1658, Latvia
grabis@rtu.lv

[2] Computer and Systems Sciences, Stockholm University, 16407 Kista, Sweden
{chenhsi.tsai,jelenaz,js}@dsv.su.se

**Abstract.** Information Systems (IS) of modern organizations and enterprises often rely on a network of partners' IS to deliver the services. The resilience of this network is the necessary condition for the operation of such ISs. The Digital Business Ecosystem (DBE) theory has emerged as an approach to ensure functioning and resilience in dynamic and open networks. This paper presents three cases of analysis of resilience of DBEs. The objective of the analysis is to assess the resilience of DBEs during its design phase. During this phase, often, only structural information presented in ISs models is available. In order to assess the resilience, the DBE models are analyzed for the potential for fulfilment of typical ecosystem goals and roles. The three DBE cases analyzed are winter road maintenance, digital vaccine, and Covid-19 testing. The paper evaluates the resilience of the DBEs and formulates the practices for uncovering and strengthening it.

**Keywords:** Information systems design · Digital business ecosystem · Resilience

## 1 Introduction

Modern information systems (IS) are not isolated and often operate in a network of partners to deliver their services. Such networks are, for example, frequently observed in mobile telecommunication industry [1] and e-commerce [2], but they are also increasingly common in other areas such as public services or healthcare. The resilience of these networks is the necessary condition for the operation of such ISs. The Digital Business Ecosystem theory has emerged as an approach to ensure resilience in dynamic and open business networks. Digital Business Ecosystem (DBE) is a virtual environment populated by digital entities such as software applications, hardware and processes [3]. Resilience is defined as the ability to remain or recover to a stable state to continuously operate during and after a crucial mishap or under constant stress [4]. Hence, the abundance of actors, or species, if we draw parallels to natural ecosystems, the interconnections among these species, and the coevolution and evolution among them, are the key factors of resilience of not only natural ecosystems but also DBEs. This leads us to the core of positioning resilience in a DBE to monitor coevolution and evolution of the

Ē. Nazaruka et al. (Eds.): BIR 2022, LNBIP 462, pp. 53–68, 2022.
https://doi.org/10.1007/978-3-031-16947-2_4

actors during and after context changes in a DBE. Furthermore, this needs to be done in conjunction with a number of overarching DBE resilience goals.

Various resilience measurements have been proposed in the literature [5, 6] and they mainly rely on the network theory or analyzed dynamics of fully instantiated representations of the ecosystem. However, this information is not available before the system is deployed and, hence, it is difficult to assess the resilience during the design time. We have identified archetypal goals [7] and roles [8], needed to be present in DBEs to ensure resilience. The presence of these goals and roles can be perceived as a structural characteristic of successful DBEs. This characteristic can be analyzed using a model-based approach, for example, Enterprise Modeling (EM) has been used to specify data ecosystems and to analyze their properties [9] and conceptual modeling has been used to connect ecosystem actors with value creation processes [10]. EM is used to represent constituent parts of the DBE such as goals and roles and these can be cross-examined with regard to the typical or reference goals and roles of the DBE. The DBEs are open and continuously evolving systems and, hence, the models are continuously updated and maintained to represent the most current information. The type of information changes during the IS development life-cycle and multiple model development phases such as design, deployment and management are distinguished. The deployment of a DBE design means aligning actor types and goals with the actual actors. This is a continuous process. Things change at the time of management but the changes are not widely known or announced. Some ways for a DBE to become non-resilient would be by some actors exiting the ecosystem, or having only some actor types engaged, or stop fulfilling some of the goals. Hence, monitoring DBEs with respect to how the various internal and external changes affect its resilience is necessary.

In summary, ISs of business ecosystems are subjected to unknown and unexpected disturbances. They require resilience, but it is difficult to evaluate it during the design time because (i) the design might be known only on the level of actor types and their general goals, (ii) the alignment between the actor types and ecosystem actors is not known or not well-elaborated, (iii) resilience and business goal alignment is not known, not done, or nor easy to establish, and (iv) decentralized or hybrid mode of DBE design.

To address these issues, we have identified the structural properties characteristic to resilient ecosystems by examining three cases according to these structural properties. This investigation is a part of an on-going research effort on development of resilient ISs relying on ecosystem services.

The objective of this exploratory paper *is to assess the structural resilience of DBEs and to identify the initial set of guidelines for designing resilient DBE systems.* Three EM cases are explored to address digital services important to citizens' well-being. The contributions of the paper are: 1) evaluation of resilience of digital business ecosystems, 2) formulation of the structural resilience index, and 3) recommendations for analysis of ecosystem resilience.

The rest of the paper is structured as follows. Section 2 reviews literature on resilience of DBEs in relation to ISs development. The analysis framework is described in Sect. 3. The case study design is discussed in Sect. 4. The analysis results and lessons learnt are presented in Sects. 5 and 6, respectively. Section 7 presents concluding remarks.

## 2   Related Work

Moore in 1993 suggested a new idea of cooperative networks which resembles an ecological ecosystem: a *business ecosystem* [11]. A *business ecosystem* bears similarity to an ecological ecosystem as being a complex system involving evolution and co-evolution. Later, the concept of *'business ecosystem'* was defined as: "an economic community supported by a foundation of interacting organizations and individuals– 'the organisms of the business world'." [12] With time, co-evolution occurs, i.e., involved organizations and individuals evolve their roles and capabilities. In recent years, the concept of ecosystem has gained awareness and significance the IS field.

A DBE captures the co-evolution between the business aspect and its partial digital representation in the ecosystem [3]. Characteristics of DBEs are digital environment, heterogeneity, symbiosis, co-evolution, and self-organization [13]. Digital environment refers to platforms or technical infrastructures where a collection of digital tools, services, other digital representations, and information can be shared and used by DBE actors to create innovations and enhance performance [13]. Heterogeneity denotes the constitution of DBEs with different features and types of actors. Symbiosis emphasizes the relationships among DBE actors that depend on each other in particular ways and get benefits or co-create greater value through the interdependencies [13]. Co-evolution refers to the collective transformation of DBE actors from one stage to another, especially their capabilities and roles, while facing opportunities and threats. Self-organization indicate DBEs' ability to learn from their environments and accordingly respond by adjusting to the changing contexts [13].

Using modelling as an approach, which aims to reduce complex domains through abstraction, can be an appropriate way for describing DBEs [14, 15]. Nevertheless, the current state of the art (reported in [16]) suggests that the area of DBE design and modelling methods lacks holistic yet feasible solutions that address the multiple aspects of a DBE and support the design and management of a DBE throughout its lifecycle. Aldea's [17] study successfully included more essential concepts (*actor, role, capability, relationship, and digital component*). Similarly, capabilities, context and goals are considered for risk analysis in financial ecosystems [18]. A preliminary data ecosystem meta-model [19] proposes to model actors, roles, relationships and resources. These studies do not elaborate methods for guiding the modeling process and analysis. A few examples of studies that suggested comprehensive methods were the Methodology of Business Ecosystem Network Analysis [20], the method for modelling interdependencies between DBE partners [21], and the approach for modelling and analysing DBEs from the value perspective [22]; still, none of them have proposed a complete method encompassing all relevant concepts, nor the methods considered both design, analysis and management of DBE. A novel contribution was seen in [23] where a top-down policy-based DBE modelling approach with its procedure was proposed. Among the proposed procedures, the most prominent steps of these procedures, such as *identifying actors, roles, or digital components*, were in accordance with the commonly included modelling elements.

## 3  Resilience Framework

The research objective is the evaluation of ecosystem resilience given the enterprise level model at its various stages of development. It is assumed that the pre-condition of the resilience is a support for archetypal goals and roles of DBE resilience, referred to as structural resilience. These goals and roles have been identified and analyzed in [7] and [8]. The roles in the DBE are: 1) *Driver* – sets the vision for the DBE and facilities its growth; 2) *Aggregator* – aggregates capabilities and resources; 3) *Modular Producer* – provides resources; 4) *Complementor* – provides resources that complement the core resources; 5) *Customer* – pays for DBE's services; 6) *End User* – uses DEB's services; 7) *Governor* – governs all actors within a DBE by providing the standards, laws etc.; and 8) *Reputation Guardian* – ensures trustworthiness.

The resilience goals in DBE are *diversity, efficiency, adaptability,* and *cohesion.* Diversity is the variety of actors for organizational units and roles, the collection of multiple resources and resource variety, and the collection of multiple capabilities and capabilities variety in a DBE. Efficiency is the resource productivity and utilization in a DBE and value delivered relative to total resource consumption. Adaptability is the transparency in terms of exposing the means of adaptation and flexibility as the ease with which a DBE can be changed. Cohesion denotes the strength of partnerships, the alignment and tightness among actors and their capabilities towards fulfilling the mission of a DBE.

**Table 1.** Modeling constructs used

| What needs to be represented | Concept | Description | Graphical symbol |
|---|---|---|---|
| Intention concerning resilience of the DBE | Resilience goals (RG) | According to [7] | Resilience goal <resilience goal name> |
| Generic role necessary for a resilient DBE | DBE resilience role (RR) | According to [8] | Resilience role <Resilience role> |
| The intention of the ecosystem | Ecosystem goal (EG) | Common goals agreed among DBE actors or Driver's vision | Ecosystem goal <ecosystem goal name> |
| Roles significant to the DBE design that organizational units can fulfill | Generic role (GR) | Specific to business domain or case | Generic role <Generic role> |
| EM goal (4EM notation) | Business goal (BG) | According to [23] | Business goal <business goal name> |
| EM organizational unit (4EM notation) | Specific organizational unit (OU) | According to [23] | Organizational unit <Organizational unit> |

The resilience of DBEs is analyzed using EM [23]. The EM technique is used to identify goals and roles specific to the cases considered and associated with the DBE resilience goals and roles. The EM process consists of three phases:

- Design phase – goals and roles are modeled at the ecosystem level without considering organizational context;
- Deployment phase – goals and roles are modeled at the company network level and the supporting IS is in development and deployment phases of its life-cycle;
- Management phase – goals and roles are detailed at the level of individual entities and the supporting IS is in the operation phase of its life-cycle.

During this progressive elaboration of the goals and roles, they are categorized as resilience, ecosystem and business goal and ecosystem and generic roles and organizational units, respectively (Table 1). The relationship between goals is defined as the ecosystem goals supporting the resilience goals and the business goals supporting the ecosystem goals. Similarly, the relationship between roles implies that the generic roles fulfill the ecosystem roles and the organization units fulfill the generic roles. The relationship between goals and roles define that certain roles are required to fulfill the goal.

## 4  Case Study Design

The ecosystem resilience is evaluated for three DBEs: Winter Road Maintenance, Digital Vaccine, and Covid-19 Testing. All three cases deal with ensuring quality of public services and well-being of citizens. They involve multiple-stakeholders and their success depends on collaboration among these stakeholders. The modeling of DBE was performed as a part of development of an IS supporting the ecosystem.

### 4.1  Research Questions and Case Selection

The following research questions are formulated: Q1: Do DBEs meet structural resilience requirements? Q2: Which aspects of resilience are under-represented in digital business ecosystems? Q3: Does contribution to achieving the resilience goals depend on the resilience role? Q4: How can resilience be assessed during the DBE design and deployment phases?

The ISs developed by companies involved in the case studies are being developed on the premise that they support and function within a resilient encompassing DBE. The first question is motivated by the need to validate this premise and to evaluate the degree of resilience in early stages of system's development. The second question aims to identify the underrepresented resilience aspects. This information is used to guide the modeling process in the case of underspecified model or to suggest required actions to improve resilience of the DBE. The resilience assessment is performed at various stages of the model elaboration and the third question explores the completeness of the resilience assessment at these stages (what do we need to know about the DBE to assess its resilience). Complementary questions are: a) Are resilience roles present in cases? b) Presence of roles contributes to ecosystem resilience, c) Do company's goals encompass resilience goals? and d) When do DBEs become non-resilient.

The cases are selected to meet requirements associated with DBEs, namely, heterogeneity, symbiosis, coevolution, self-organization. A description on these characteristics of DBEs serving as requirements is given in Sect. 2.

### 4.2  Case Descriptions

*Winter Road Maintenance (WRM)*

This research is carried out as a part of the industry sponsored project on development of an integrated ERP system for winter road maintenance (WRM). The IT consulting company provides a specialized ERP system to municipalities in Latvia. The system consists of various services including financial accounting, assets management, social welfare services and many other services. In order to expand in the area of data driven services including sensing technologies, the WRM case was selected as a pilot for developing modules of the ERP system with advanced data analytical functionality.

WRM is performed at state or regional level. This research initially focuses on the regional level, where municipalities are tasked with road maintenance and they contract service providers to maintain the roads. Both parties need to exchange information to plan, report, control and analyze the WRM operations. These operations strongly depend on external factors [24] obtaining data about road conditions. A number of parties including citizens, public transportation and emergency services need up-date information on road conditions and maintenance work. Thus, the case is characterized by rich information flows among the actors.

*Digital Vaccine (DV)*

The digital vaccine case used as an example for this study started as a European Institute of Innovation and Technology Health (EIT Health) Innovation by Design project [25]. The case is based in Stockholm, Sweden and has also been reported in [7, 8]. The digital vaccine DBE consists of actors such as digital and physical health service providers, health product suppliers, the digital vaccine company (Health Integrator AB, HI), citizens as individual users, public sector (Stockholm region), and investors. Health Integrator, as one of the central actors holding the leadership role in the DBE, owns a digital health platform. The aim is to shift the focus in healthcare from reactive to proactive by providing tailored services based on personal needs and supporting healthier lifestyle habits through the digital business ecosystem. Also, a health outcomes-based contracting model is investigated by HI, Stockholm regional council, and investors.

*Covid-19 Testing (CT)*

The case is based in Stockholm Sweden as final year thesis project within a recently established company Noviral [26]. Its business is to market rapid Covid-19 tests to all actors involved performing Covid-19 testing in Sweden. It also shares the test data with researchers and Swedish Public Health Authorities to increase the knowledge of Covid-19 through a digital platform. The objective of the case was to map the DBE actors and to analyze resilience goals of Noviral.

## 4.3 Data Collection and Analysis Procedure

Modeling was performed as a part of university-industry collaboration projects on DBE development. Joint modeling sessions were organized and researchers were driving the sessions. The goal and role models were created according to the following steps:

- Initial exploration – the researchers identified ecosystem's stakeholders and prepared the guiding questions for modeling session as well as gathered information about stakeholders, which are not directly involved in the modeling process.
- Interview – the stakeholders giving information individually were interviewed and their information was structured by the analysts.
- Model development – joint sessions were organized with the stakeholders and goal and actor models were elaborated including the information gathered in the previous two steps. The following steps are performed in the design phase: a) ecosystem goals and generic roles are identified; b) the ecosystem goals are associated with the resilience goals; c) the generic roles are associated with the resilience roles and d) relationships among the resilience roles and the ecosystem goals are specified.
- The models are further refined in the deployment phase – specific organizational units are identified to fulfil the generic roles, business goals specific to the generic roles and the organizational units are also identified as well as relation among the goals and roles are updated.
- The initial resilience analysis is performed – several resilience indicators are defined and are evaluated to determine the structural resilience of the DBE.
- Model validation – the modeling and analysis results are discussed with the stakeholders to validate the ecosystem resilience assessment.

The enterprise models are analyzed to assess the DBE resilience using the information available during the design and deployment phase. Similarly as in [27], the enterprise model is perceived as a graph, where goals and roles are nodes and relations among them are represented as edges. Node degree indicators are calculated for specific elements of the model to quantify their role in the ecosystem model. Paths connecting goals and roles are identified to characterize indirect relations among them. The design phase resilience goal support indicator $I_{rg}^D$ is calculated for every resilience goal:

$$I_{rg}^D = d_{rg}^+ N^{-1}, rg \in \textbf{RG}, \tag{1}$$

where $N = \sum_{i \in \text{RG}} d_i^+$ and $d_i^+$ is in-degree for a resilience goal $rg$ and $\textbf{RG}$ is a set of all resilience goals. The indicator shows whether the resilience goals are support by the ecosystem goals and highlights the resilience goals having the largest relative support. The design phase resilience role fulfilment index $I_{RR} = |\textbf{RR}|^{-1} \sum_{i \in \text{RR}} d_i^+$ shows whether the resilience roles are supported by generic roles (i.e., the resilience roles have incoming edges), where $\textbf{RR}$ is a set of the resilience roles.

The deployment phase resilience goal support indicator $I_{rg}^L$ is calculated the same way as $I_{rg}^D$ and incoming edges are originating from both the ecosystem and business goals. The model is also analyzed to identify relationships among the resilience roles and the resilience goals. The indicator represents a count of paths leading from resilience roles to resilience goals $I^D_{(rr,rg)} = |\mathbf{P}_{(rr,rg)}| \times |\mathbf{P}|^{-1}$, where $|\mathbf{P}_{(rr,rg)}|$ is a set of unique paths $p$ leading from a resilience role $rr$ to a resilience goal $rg$ via generic roles $eg$ and ecosystem goals $rg$

$$p_i = (rr, gr, eg, rg), rr \in \mathbf{RR}, gr \in \mathbf{GR}, eg \in \mathbf{EG}, rg \in \mathbf{RG}. \tag{2}$$

The same indicator can be calculated in the deployment phase model and paths all could lead from a resilience role to a resilience goal via business goals.

## 5   Results

*Winter Road Maintenance*
The ecosystem goals were identified jointly by representatives of various stakeholders involved in the WRM case. These goals cover the whole road maintenance life-cycle from monitoring, road clearing, control, feedback and analysis (Fig. 1). They also represent well-being, cost and environmental concerns. To clear roads goal is associated with efficiency. To create new services goal contributes to both diversity (new capabilities are added to the ecosystem) and adaptability (replaceable services are available). The support for the Cohesion goal was not immediately obvious. To control road maintenance supports cohesion because it enforces compliance with the regulatory requirements concerning road maintenance and To provide environment monitoring improves transparency contributing to involvement of various stakeholders.

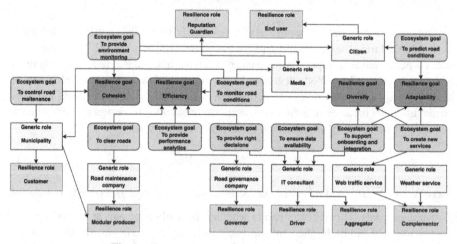

**Fig. 1.** The WRM ecosystem design stage model.

Similarly, the generic roles are identified. The road maintenance platform is developed by an IT consulting company initiating collaboration in the ecosystem. Thus, the consulting company services are both Driver and Aggregator. The road maintenance services are provided by a multitude of road maintenance companies and there is a

nation-wide road governance company, which sets maintenance requirements and provides primary infrastructure. The Customer role is fulfilled by municipalities contracting
the road maintenance companies and using the platform services provided by the IT consultant. The municipality and road maintenance company also provide various services
to the ecosystem. The ecosystem also involves several types of complementors though
they do not have obvious goals.

The Media and Citizen roles are identified without actually involving any stakeholders representing these roles. They were mentioned by municipalities as driving forces
behind improving efficiency and transparency of maintenance activities. The road governance company was approached only in separate interviews. Even though they set
road maintenance requirements and provide oversight and mediation. The stakeholders
did not emphasize these aspects and they are not represented in the DBE model. The
company itself focused on the requirements side. There is potentially a large number of
data providers deemed as Complementors. Their involvement was also uncovered indirectly by other stakeholders and they do not appear as conscious members of the DBE.
Ten ecosystem goals and eight generic roles were identified during the design phase. It
is determined that all resilience goals are supported while the Efficiency goal has the
largest support and the Cohesion goal has the lowest support (Fig. 2). All resilience roles
are also supported. The average number of edges per generic role is 1.375 and the Driver
and Aggregator resilience roles are supported by the generic role IT consultant.

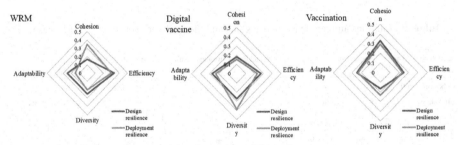

**Fig. 2.** Resilience goal support indicators $I_{rg}^D$ and $I_{rg}^L$ for the cases considered.

The design model highlighted significant aspects for further exploration. The additional information was represented in the deployment phase model including adding
specific goals (Fig. 3). The relationships are also established among the elements following the principle that new relationship is added only if it cannot be derived using
existing relationships (e.g., a business goal to resilience goal relation is added if there
was no relation between the resilience goal and an ecosystem goal associated with the
business goal). The Cohesion goal becomes the most important after adding business
goals (Fig. 2). There are 8 direct relationships among the business goals and resilience
goals established in the deployment phase. The radar charts indicated that the initial
ecosystem model was skewed towards efficiency and much of the modeling emphasis
has been diverted to cohesion once the stakeholders have realized importance of the
ecosystem.

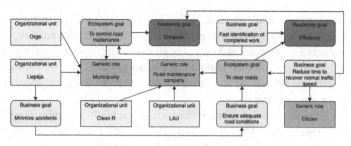

**Fig. 3.** A fragment of the deployment stage ecosystem model.

The relations among resilience goals and the resilience roles are derived (Table 2). Most of the roles contribute to the efficiency and diversity goals. The End user is the most significant contributor while none of the roles alone contributes to all resilience goals. The Cohesion goal is supported only by four roles and surprisingly the Driver and the Aggregator are not among them at the design phase. This aspect was elaborated during the deployment phase. As a result, $I_{(rr,rg)}$ *(deploy)* shows increasing support for the Cohesion goals in the ecosystem. The Efficiency goal appears to be the most important to the stakeholders. The End user role remained the most important contributor at the deployment stage as well while the importance of the Driver significantly increased as the business goals were considered.

**Table 2.** Relations among the resilience goals and the resilience roles in the WRM case.

| Resilience goal | Driver | Aggregator | Modular Producer | Complementor | Customer | End user | Governor | Reputation guardian | $I_{(rr,rg)}$ *(design)* | $I_{(rr,rg)}$ *(deploy)* |
|---|---|---|---|---|---|---|---|---|---|---|
| Cohesion | 0 | 0 | 1 | 0 | 1 | 1 | 0 | 1 | 0.18 | 0.26 |
| Efficiency | 1 | 1 | 1 | 0 | 1 | 1 | 1 | 0 | 0.27 | 0.35 |
| Diversity | 1 | 1 | 1 | 1 | 0 | 2 | 0 | 1 | 0.32 | 0.24 |
| Adaptability | 1 | 1 | 0 | 1 | 0 | 1 | 0 | 1 | 0.23 | 0.15 |

*Digital Vaccine*
The DBE and business goals for the digital vaccine case were reported in [7] and elaborated in this study with stakeholders. The design stage analysis revealed that there is a linear alignment between the resilience goals cohesion and efficiency and their corresponding ecosystem goals, meaning that each of these ecosystem goals exclusively contributes to a single resilience goal. For the Diversity and Adaptability goals, the contributing ecosystem goals are more overlapping. The Efficiency and Adaptability goals are each supported by three ecosystem goals, whereas the Diversity and Cohesion goals are supported by two ecosystem goals (Fig. 4). All resilience roles are supported, with the Driver, Aggregator and Reputation guardian resilience roles being supported by the generic role health platform provider.

The deployment phase analysis revealed an emphasis on diversity and adaptability. The information system developed is a portal and the driver was keen on expanding the offering a wide range of products and services and the flexibility of them.

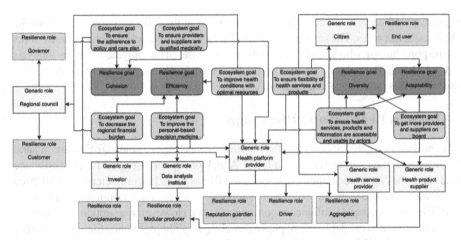

**Fig. 4.** The Digital Vaccine ecosystem design stage model.

Most of the roles contribute to the efficiency goals (Table 3). The Driver, Aggregator, and Reputation guardian (played by the digital vaccine company, HI) are the most significant contributors to all resilience goals in this case. The Diversity aspect was emphasized during the deployment phase and elaborated with business goals by some contributing actors, resulting in increased $I_{(rr,rg)}$ *(deploy)* indicators for this aspect. The Diversity goal appears to be the most important as suggested by the $I_{(rr,rg)}$ *(deploy)* indicators. An observation during the deployment phase is that some of these resilience goals (and there contributing ecosystem goals) are supported by generic roles without the actual organizational units being known. This results in a situation where some of the ecosystem goals are not supported by identifiable business goals.

**Table 3.** Relations among the resilience goals and the resilience roles in Digital Vaccine.

| Resilience goal | Driver | Aggregator | Modular Producer | Complementor | Customer | End user | Governor | Reputation guardian | $I_{(rr,rg)}$ *(design)* | $I_{(rr,rg)}$ *(deploy)* |
|---|---|---|---|---|---|---|---|---|---|---|
| Cohesion | 2 | 2 | 0 | 0 | 1 | 0 | 1 | 2 | 0.16 | 0.14 |
| Efficiency | 3 | 3 | 1 | 1 | 2 | 0 | 2 | 3 | 0.30 | 0.27 |
| Diversity | 2 | 2 | 4 | 0 | 0 | 1 | 0 | 2 | 0.22 | 0.30 |
| Adaptability | 3 | 3 | 6 | 0 | 0 | 1 | 0 | 3 | 0.32 | 0.29 |

## Covid-19 Testing

DBE model of Noviral was elaborated in a top-down approach having the association between the resilience and the ecosystem goals already in mind during the modeling session. That resulted in linear relationships among these goals and in quite uniform distribution of the support for the resilience goals. The further elaboration in the design phase uncovered additional contributions from the business goals. Particularly, the diversity aspect was emphasized because delivery of many services and testing products depends on strong collaboration among the actors in the DBE. The actor model represented the organizational units involved in being generic roles in the DBE and how they fulfill the DBE resilience roles. The actor and goal modeling led to conclude which resilience roles are responsible for which resilience goals. In this case, the deployment phases analysis did not change the balance among the resilience goals. Due to lack of space enterprise models of this case have been omitted, c.f. [26] for details.

## 6  Lessons Learned and Discussion

The analysis of resilience goals and the resilience roles helps to guide DBE modeling and to understand the characteristics of a DBE. A lack of supporting goals for the resilience goals and roles indicated either weaknesses in DBE or model incompleteness. In all three cases, the resilience goals and roles were supported already at the design stage indicating that DBE met structural resilience requirements (Q1). However, the support was uneven, which increased in the deployment phase revealing implicit preference for one or another aspect of the DBE resilience. The DBE modeling was unavoidably influenced by the stakeholders present during the modeling sessions and interviews. Using the resilience goals and roles for guiding the modeling helped to identify other relevant DBE roles and participants, which could have been overlooked following a more traditional EM process of goal and actor discovery and modeling.

The DBEs of WRM and CT focused on efficiency and cohesion while adaptability and diversity were under-represented. The Digital Vaccine case, in contrast, focused heavily on diversity (Q2). Concerning resilience roles, the customer, complementor, and governor were under-represented even if they appear to have a crucial role to achieve the resilience goals. This can be explained by difficulties to involve stakeholders representing these roles in the modeling sessions. This group is also the most open group of stakeholders. The WRM case was analyzed from a more holistic perspective but for the cases of DV and CT the models were created from the point of view of the DBE drivers. The drivers, and other roles, should ensure that appropriate processes are established to involve DBE stakeholders in roles such as complementor, governor, and reputation guardian. These processes should be in place during the management phase and they should be adaptive. This would lead to the increasing importance of the Adaptability goal, which was under-represented suggesting that the DBE is somewhat restricted and has no clear vision to deal with the open and dynamic nature of DBEs.

Drivers, aggregators, and customers are drawn to the DBE primarily because of resilience goals (Q3). In doing so, they clearly realize the importance of cohesion. It would be expected that cohesion is tightly associated with governors and reputation guardians. However, they were relatively passive members of the DBEs and the latter were involved mostly indirectly. In the CT case, the governor issued the specifications for the products and received testing data. The role of customer was significant and well-elaborated in all three DBEs. In the process of analysis, information about customer and complementors can be gathered from various sources including public sources as it was done for the cases of Digital Vaccine and Covid-19 Testing.

Analysis during the design and deployment phase clearly indicated whether all reference goals and roles concerning resilience are considered in the DBE structure (Q4). The model-based analysis at the design phase does not allow to measure resilience quantitatively. It tests the necessary pre-conditions for resilience though says little about the sufficient conditions. However, the evolution of enterprise models starting from design, deployment, and then to management allows the identification of critical aspects of ensuring resilience. The biggest advantage of the resilience analysis is providing guidance for EM to pinpoint aspects that need to be discovered and analysed.

The main lessons learned from the modeling process are that DBE drivers are often unaware of the extent of the DBE and the reference to resilience goals and roles helps to extend the discussion and to strengthen their understanding. They acknowledged that one could hardly speak about a DBE if the resilience goals and roles are not considered.

We also observed that there is a significant overlap between the driver and aggregator roles in DBE. Our cases show some evidence that these roles should be separated. The complementor role can be implemented by several generic roles and organizational units. The impact of this role is difficult to assess using structural analysis. However, the complementors often have played significant roles in successful DBEs.

Modeling of the cases reported here faced the challenge of how to perform modeling to cover the important aspects of DBE. This was solved by experimentations and discussions among modelers. In the future this process can be sped up by DBE resilience modeling patterns. For example, three candidate patterns have been identified in the cases reported in this paper:

- Pattern 1: there is an ecosystem goal and no corresponding business goal: in this situation each actor should accept this ecosystem goal to become its business goal; the business goal itself could be refined to the context of the particular actor.
- Pattern 2: there is an ecosystem goal, and business goals are present: in this situation the ecosystem goals and business goals are compared, and if there is a conflict, either a) the actor accepts the ecosystem goal by negotiation, b) the actors upon negotiation decides on a new ecosystem goal, or c) when the actors cannot agree, the driver helps to making the final decision about the ecosystem goal that will be then propagated to the business goals.
- Pattern 3: there are business goals but no a corresponding ecosystem goal: in this situation, the actor of a relevant business goal for the DBE, or several actors sharing a same business goal, propose it to be considered on the ecosystem goal level; the business goal would need to be revisited and generalized.

There are several research limitations. The three cases concern DBEs in early stages of their development and the results cannot be readily applied to more mature DBEs. The indicators can only be used in relative terms within a single case because the DBEs do not have well-defined limits and neither all goals nor organizational units can be identified in the design phase. The stakeholders were involved in various ways because participatory modeling sessions involving all stakeholders are not possible in the case of DBE. It appears that the DBE design was influenced by direct participants more significantly than by interviewed and indirect stakeholders. Finally, the management phase concerning live operations of the DBE was not considered because the DBEs are not fully functional as of now. Cross-examination of the conclusions drawn in the design and management phases with live observation in the management phase would provide empirical sources for additional validation.

# 7  Conclusions

The current literature suggests that design time evaluation of resilience of DBEs and their supporting ISs is challenging. Therefore, it was proposed to use structural analysis of enterprise models to assess the resilience in early stages. The structural resilience was evaluated with regard to the reference resilience goals and roles identified in DBE research. Three cases were analyzed to assess the potential and challenges of such an approach. Analysis of the cases confirms that the resilience goals and roles are essential to facilitate the discussion on resilience of DBEs. The DBEs analyzed exhibit the pre-conditions for achieving resilience and the analysis improved stakeholder understanding of the ways to improve resilience of their DBEs. The analysis also helps in validating the set of resilience goals and roles since they are readily recognized by the stakeholders. However, the ecosystem model and emerging IS are still shaped by the main stakeholders, serving as drivers or aggregators and, hence, a combination of various techniques is needed to improve representation of other roles such as complementors and customers who have a major impact but are under-represented in design activities.

This paper focused on the goals and roles because they can be readily evaluated against the generic resilience roles and ecosystem goals. This is just the first step and the resilience evaluation should also analyze the dynamic features, e.g., by looking at the process perspective. However, processes are more domain specific and there are few known reference processes supporting the development of resilient systems. Currently, the resilience evaluation requires the development of dynamic process models such as simulation models.

The tool support for continuous elaboration and maintenance as well as conducting analysis is required. It can be developed by extending suitable meta-modeling platform and utilization of tools for graph analytics [28, 29].

# References

1. Basole, R.C.: Visualization of interfirm relations in a converging mobile ecosystem. J. Inf. Technol. **24**, 1–16 (2009)
2. Aulkemeier, F., Paramartha, M.A., Iacob, M.-E., van Hillegersberg, J.: A pluggable service platform architecture for e-commerce. IseB **14**(3), 469–489 (2015). https://doi.org/10.1007/s10257-015-0291-6
3. Nachira, F., Dini, P., Nicolai, A.: A network of digital business ecosystems for Europe: roots, processes and perspectives. In: Nachira, F. et al. (eds.) Digital Business Ecosystems. European Commission, Bruxelles (2007)
4. Wreathall, J.: Properties of resilient organizations: an initial view. In: Hollnagel, E., et al. (eds.) Resilience Engineering: Concepts and Precepts. Ashgate (2006)
5. Basole, R.C., Russell, M.G., Huhtamaki, J., Rubens, N., Still, K., Park, H.: Understanding business ecosystem dynamics: a data-driven approach. ACM Trans. Manag. Inform. Syst. **6**(2), Article 6 (2015)
6. Wang, J.W., Gao, F., Ip, W.H.: Measurement of resilience and its application to enterprise information systems. Enterp. Inf. Syst. **4**(2), 215–223 (2010)

7. Tsai, C.H., Zdravkovic, J., Stirna, J.: Capability management of digital business ecosystems – a case of resilience modeling in the healthcare domain. In: Herbaut, N., La Rosa, M. (eds.) CAiSE 2020. LNBIP, vol. 386, pp. 126–137. Springer, Cham (2020). https://doi.org/10.1007/978-3-030-58135-0_11

8. Tsai, C.H., Zdravkovic, J.: A survey of roles and responsibilities in digital business ecosystems. In: Proceedings of PoEM'2020 Forum: CEUR-WS.org (2020)

9. Kampars, J., Zdravkovic, J., Stirna, J., Grabis, J.: Extending organizational capabilities with open data to support sustainable and dynamic business ecosystems. Softw. Syst. Model. **19**(2), 371–398 (2019). https://doi.org/10.1007/s10270-019-00756-7

10. Kaiser, C., Stocker, A., Viscusi, G., et al.: Conceptualising value creation in data-driven services: the case of vehicle data. Int. J. Inf. Manag. **59**, 102335 (2021)

11. Moore, J.F.: Predators and prey: a new ecology of competition. Harv. Bus. Rev. **71**(3), 75–86 (1993)

12. Moore, J.F.: The Death of Competition: Leadership and Strategy in the Age of Business Ecosystems. Harper Business, New York (1996)

13. Senyo, P.K., Liu, K., Effah, J.: Unpacking the role of political-will in digital business ecosystem development for socioeconomic benefits. In: Proceedings of ECIS 2019. AIS (2019)

14. Wieringa, R., Engelsman, W., Gordijn, J., Ionita, D.: A business ecosystem architecture modeling framework. In: Proceedings of CBI 2019. IEEE (2019)

15. S Oliveira, M.I., Barros Lima, G.D.F., Farias Lóscio, B.: Investigations about data ecosystems: a systematic mapping study. Knowl. Inf. Syst. **61**, 589–630 (2019)

16. Tsai, C.H., Zdravkovic, J., Stirna, J.: Modeling digital business ecosystems: a systematic literature review. Complex Syst. Inf. Model. Q. **30**, 1–30 (2022)

17. Aldea, A., Kusumaningrum, M.C., Iacob, M.E., Daneva, M.: Modeling and analyzing digital business ecosystems: an approach and evaluation. In: Proceedings of CBI 2018. IEEE (2018)

18. Feltus, C., Grandry, E., Fontaine, F.-X.: Capability-driven design of business service ecosystem to support risk governance in regulatory ecosystems. CSIMQ (10), 75–99 (2017)

19. Oliveira, M.I.S., Oliveira, L.E.R.A., Batista, M.G.R., Loscio, B.F.: Towards a meta-model for data ecosystems. In: Proceedings of DG.O 2018. ACM (2018)

20. Battistella, C., Colucci, K., De Toni, A.F., Nonino, F.: Methodology of business ecosystems network analysis: a field study in Telecom Italia future centre. Technol. Forecast. Soc. Change **80**(6), 1194–1210 (2013)

21. Senyo, P.K., Liu, K., Effah, J.: Towards a methodology for modelling interdependencies between partners in digital business ecosystems. In: Proceedings of LISS 2017. IEEE (2017)

22. Biermann, J., Corubolo, F., Eggers, A., Waddington, S.: An ontology supporting planning, analysis, and simulation of evolving digital ecosystems. In: Proceedings of MEDES 2016, pp. 26–33. ACM (2016)

23. Sandkuhl, K., Stirna, J.: Capability Management in Digital Enterprises. Springer, Cham (2018). https://doi.org/10.1007/978-3-319-90424-5

24. Dey, K.C., et al.: Potential of intelligent transportation systems in mitigating adverse weather impacts on road mobility: a review. IEEE Trans. Intell. Transp. Syst. **16**, 3 (2015)

25. EIT Health. https://eithealth.eu/project/the-health-movement/. Accessed 10 Apr 2020

26. Linde, C., Pedersen, K.: Analysis of a digital business ecosystem of a rapid test distributor. Swedish, Department of Computer and Systems Science, Stockholm University (2021)

27. Smajevic, M., Bork, D.: Towards graph-based analysis of enterprise architecture models. In: Ghose, A., Horkoff, J., Silva Souza, V.E., Parsons, J., Evermann, J. (eds.) ER 2021. LNCS, vol. 13011, pp. 199–209. Springer, Cham (2021). https://doi.org/10.1007/978-3-030-89022-3_17

28. Schoder, M.-S., Utz, W.: BEDe: a modelling tool for business ecosystems design with ADOxx. In: Camarinha-Matos, L.M., Boucher, X., Afsarmanesh, H. (eds.) PRO-VE 2021. IAICT, vol. 629, pp. 526–535. Springer, Cham (2021). https://doi.org/10.1007/978-3-030-85969-5_49

29. Grabis, J., Deksne, L., Roponena, E., Stirna, J.: A capability-based method for modeling resilient data ecosystems. In: Karagiannis, D., Lee, M., Hinkelmann, K., Utz, W. (eds.) Domain-Specific Conceptual Modeling. Springer, Cham (2022). https://doi.org/10.1007/978-3-030-93547-4_15

# A Case Study on *its*VALUE to Evaluate its Method, Notation and ADOxx Modeller

Henning D. Richter[✉] [iD]

University of Rostock, 18051 Rostock, Germany
`henning.richter@uni-rostock.de`

**Abstract.** The new ITIL 4 standard was introduced in 2020. Still, ITIL standardization highly influences how IT-Service Management (ITSM) is seen and performed in practice. Hence, this new standard may have a similarly high impact. The major focus on StakeholderValue (SV) is a key element of ITIL 4 for analysing IT-Services. Generally, stakeholder orientation is a current trend in business analysis. Therefore, our *its*VALUE method and modeller provide means to model and analyze value delivery in IT-Services. In an ITIL 4 context, both can also be used for the design of services. *its*VALUE combines "traditional" approaches to Value Stream (VS) analysis and service modelling. Further, it adds concepts and functionalities that meet the requirements of ITSM and ITIL 4. Still, *its*VALUE must be checked in practice regarding its method, notation and ADOxx modeller. To investigate these issues, this work deeply describes and discusses a first case study we carried out regardingly. Overall, the participants involved perceived the *its*VALUE method as suitable to support value-oriented ITSM. However, there were slight differences in their perception of the *its*VALUE notation. Lastly, they perceived the *its*VALUE ADOxx modeller as a powerful and supportive tool.

**Keywords:** ITIL 4 · Service modelling · Service value · Value stream modelling · Stakeholder value · Service blueprinting · IT-Service Management · Case study · Method evaluation · Notation evaluation · ADOxx modeller · Prototype evaluation

## 1 Introduction

In the past years, services have become the biggest and most dynamic market component of both industrial and developing countries. Further, they are the most important goods for generating organizational value; for both the company itself and their customers. Additionally, almost any current service is supported by IT components, while IT itself is developing as fast as never before. Thus, companies can benefit from enhancing their understanding and performance in IT-Service Management (ITSM). New techniques (e.g., cloud computing, machine learning, blockchain, etc.) enabled new opportunities for the value

© The Author(s), under exclusive license to Springer Nature Switzerland AG 2022
Ē. Nazaruka et al. (Eds.): BIR 2022, LNBIP 462, pp. 69–85, 2022.
https://doi.org/10.1007/978-3-031-16947-2_5

chains and value creation of companies. Hence, IT (especially ITSM) is one of the most important business drivers companies should carefully consider nowadays to achieve competitive advantages [1–3].

For ITSM, ITIL v3 is a well-known reference for best practices among practitioners and scientists. It describes processes, roles and Key Performance Indicators (KPIs). But a revision was necessary due to current trends (e.g., increasing market dynamics, the advent of agile software development and the integration of products and services). To address these issues, ITIL 4 was released in 2020. It primarily focuses on enabling companies to respond to new stakeholder demand quickly and simply. It also states that a company's purpose is to create value for its stakeholders. In this regard, everything a company does must serve (directly or implicitly) the creation of value for their stakeholders. As ITIL itself has a strong industrial background, it is likely that many companies will adopt the new ITIL 4 to improve their ITSM capabilities. While ITIL 4 describes these capabilities and their integration generally, a concrete method or tool-set for the integration of stakeholder value in service design is not provided. Even if an enterprise does not intend to implement ITIL 4, considering Stakeholder Value (SV) in IT-Service Management can improve demand orientation [1–5].

To begin with, we performed a Structured Literature Analysis (SLA) to collect approaches suitable for ITIL 4. With our SLA [6], we found approaches like IT Self-Service Blueprint (IT S-SB) by Bär et al. [7] or Value Stream Method 4.0 (VSM 4.0) by Hartmann et al. [8,9]. These approaches support modelling and analysis of IT-Service delivery from a value and stakeholder-oriented perspective. However, all approaches we discovered are either specialized on a certain use case (e.g., IT S-SB) or just miss some aspects and requirements of value delivery modelling that are important in ITIL 4. Further, notations like Value Delivery Modelling Language (VDML)[1] and the Archimate Motivation Extension[2] provide necessary modelling concepts in this regard. Unfortunately, no method support exists in terms of procedures and guidelines for the creation and usage of such models. As these notations remain on a high abstraction level, their usage for value-oriented ITSM would imply a further operationalization as well.

To address this gap, we developed "It's a Value Added Language You Employ" (*its*VALUE). It consists of a method, notation and tool (the *its*VALUE modeller). Further, it serves modelling and analysing value delivery in ITSM. We build it on proven concepts of service, value and enterprise modelling. In Sect. 2, we briefly introduce *its*VALUE and 1) its method components and its general process of application, 2) its notation forming the base for our modelling tool implementation and 3) our ADOxx Modeller. To evaluate these three artefacts, we carried out a case study deeply described in Sect. 3. In Sect. 4, we introduce our corresponding results followed by a discussion in Sect. 5. Then, Sect. 6 outlines our limitations. Lastly Sect. 7 describes the current development state and provides an outlook to next steps and foster *its*VALUE application.

---

[1] https://www.omg.org/spec/VDML/.

[2] https://pubs.opengroup.org/architecture/archimate3-doc/.

# 2    *its*VALUE: A Method, Notation and ADOxx Modeller

In this paper, we can describe *its*VALUE only very briefly due to limited space. However, we already published two works [10,11] more deeply addressing it. Thus, we recommend reading them, if you desire to become more familiar with *its*VALUE itself.

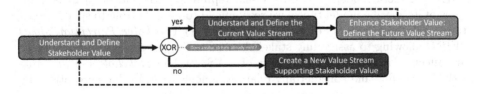

**Fig. 1.** The *its*VALUE Framework.

## 2.1    The *its*VALUE Method

As previously mentioned, *its*VALUE combines and amends proven concepts of Service-, Value-, and Enterprise Modelling to support VS analysis for IT-Services with a special focus on ITIL 4 because of its practical relevance. According to the ITIL 4 documentation [1–5], *"value"* is a set of a perceived usefulness, importance and benefits of something. This goes beyond "traditional" VS analysis and value modelling, where VS optimization is focused on processing times and modelling the exchange of economic value. We developed *its*VALUE to provide a sound combination of new ideas and requirements for value-oriented IT-Service Management based on ITIL 4 and the named "traditional" approaches. Thus, we took key elements of *its*VALUE from VSA 4.0 by Meudt et al. [12,13], VSM 4.0 and VSD 4.0 by Hartmann et al. [8,9] and VSMN by Heger et al. [14] (as extensions of "traditional" VS modelling considering information processing and stakeholder perspectives), IT S-SB by Bär et al. [7] (as an approach adding information technology to "traditional" Service Blueprints) and 4EM by Sandkuhl et al. [15] (as a participatory enterprise modelling method that supports the integration of stakeholders and provides concepts that allow modelling context influence on value delivery). Our *its*VALUE method consists of four components describing the steps to model, analyse, and (re-)design VSs of IT-Services. Figure 1 shows this method framework aligning those components in a process. Two alternatives are distinguished depending on whether there is already an existing VS or not. Further, the method should follow an iterative approach to develop the required insight into the analysed VS. As Geist [16] suggests performing four iterations with stakeholders in a modelling context, we recommend four iterations for *its*VALUE. Besides explicitly covering any value perception by the stakeholders, *its*VALUE also considers the *Digitization Rate (DR)*, *Data Availability (DA)* and *Data Usage (DU)* by Meudt et al. [13] for each process of a VS.

## 2.2  The *its*VALUE Notation

As we developed the *its*VALUE notation by a combination of existing and scientifically well elaborated notations, it should match the nine design principles by Moody [17]: *1) perceptual discriminability, 2) visual expressiveness, 3) semantic transparency, 4) dual coding, 5) graphic economy, 6) semiotic clarity, 7) complexity management, 8) cognitive integration and 9) cognitive fit.* For the *its*VALUE meta models, all of its visual representations and detailed descriptions of its specific model types, we recommend to read [11] due to space limitations. *its*VALUE consists of three model types: 1) The *Value Perception Model (VPM)* allowing to assign any stakeholder its value perceptions and how those are affected by any component part of a VS, 2) The *Stakeholder Value Map (SVM)* to define and consider any relations among several value perceptions by

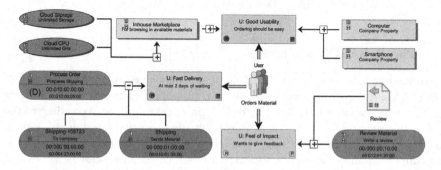

**Fig. 2.** An Exemplary Value Perception Model (VPM) for a stakeholder group called "User".

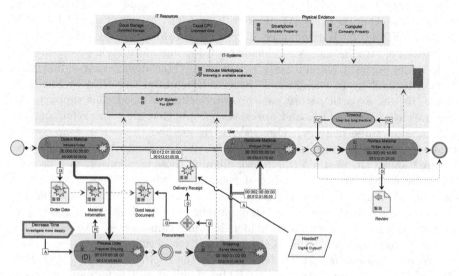

**Fig. 3.** An Exemplary Value Stream Blueprint (VSB) for a fictional service.

any stakeholder and 3) The *Value Stream Blueprint (VSB)* to visualize any VS with all components part of it (e.g., processes, IT resources, IT systems, information, waiting times, information and material flows, etc.). Figure 2 and 3 illustrate an exemplary case. In terms of better readability, this case is fictional and not originated from the case study later discussed. Very briefly summarized, Values are green, Processes either red (information related) or blue (material related), IT-systems pink, IT-Resources orange, Physical Evidences purple and lanes grey (please see [11] for more details).

### 2.3   The *its*VALUE ADOxx Modeller

In addition to the *its*VALUE method and notation, we have developed a modeller using ADOxx[3], as it consists of all required features a meta modelling platform should provide according to Karagiannis et al. [18]. As a special feature, we programmed a dynamic modelling environment where any instance of an object can exist across multiple models while staying updated everywhere in real time. Additionally, we developed four callable procedures supporting the users in VS analysis: 1) The *"Cumulative Time Calculator"* calculating cumulative times of processes and waiting times between processes inside a VS, 2) *"Receive Value Information"* automatically extracting and adding information from VPMs to a VSB, 3) The *"Data Information Calculator* calculating the DR, DA and DU of a process inside a VS and 4) The *"Waste Detector"* detecting potential spots of waste inside a VS based on pre-defined patterns.

## 3   Case Study Design

To evaluate *its*VALUE, we carried out a case study (as recommended by Baxter et al. [19] and Zucker [20] for an exploratory setting such as ours) including a modelling session followed by Guided Interviews (GIs) with the participants. We divided the interviews into four distinct parts to differentiate between our method artefacts and how they were perceived. With [10], we previously introduced a fraction of this case study only focusing on the *its*VALUE method. However, we have also investigated the *its*VALUE notation and ADOxx modeller. Further, we have evaluated the execution of our modelling session as well, as this setting may have affected the participants' perception on our artefacts somehow. Thus, we now refer to [10] for the *its*VALUE method and focus here just on its notation, ADOxx Modeller and the execution of our modelling sessions. In the following subsections, we describe our case study setting, relevant related work and how we derived both our Research Questions (RQs) and Guided Interview Questions (GIQs)

### 3.1   The Case Study Setting and Modelling Session Execution

We deeply described the content of this section already in [10]. Thus and due to limited space, we now refer to that paper [10] for an extended description and

---

[3] www.adoxx.org.

**Table 1.** Mapping the participants (P) to the groups of stakeholders involved.

| Group of Stakeholders | P A | P B | P C | P D | $\sum$ |
|---|---|---|---|---|---|
| User (Employee) | *Yes* | *Yes* | *Yes* | *Yes* | 4 |
| Global Service Desk | *No* | *No* | *No* | *No* | 0 |
| Team Leader | *Yes* | *No* | *No* | *No* | 1 |
| Client Service Team | *No* | *No* | *Yes* | *No* | 1 |
| Field Service (external) | *No* | *No* | *No* | *No* | 0 |
| Service Manager | *No* | *Yes* | *No* | *Yes* | 2 |

just summarize its most important points. We could encourage four employees to participate from the company *Drägerwerk AG & Co. KGaA*, an enterprise located in Lübeck (Schleswig-Holstein, Germany) whose core business model is the development and maintenance of technical, medical and safety equipment all over the world[4]. We investigated the VS for in-house notebook ordering. Three participants claimed having modelling experiences in advance. None of them has any relationship to the author. To provide the participants with a proper understanding on what *its*VALUE is and how it should work, we performed a participatory modelling session in advance lasting 2:30 h. Therefore, we roughly followed the framework of *its*VALUE (see Fig. 1). However, we had to do it remotely via Microsoft Teams due to the COVID-19 pandemic. Further, ADOxx does not support multiple users in parallel. Thus, we had to split the tasks assigned. Whereas the participants just discussed and shared their input, a host was guiding the discussion and progress of the modelling session. Additionally, that host was modelling the participants' input. In real time, the host's screen was shared for the entire duration of the modelling session. Due to limitations, we could only perform a single iteration of *its*VALUE on an already existing and rather small VS. Thus, we requested the participants to provide some preparations in advance (What do they value? What not? Etc.). Based on their input, we prepared first drafts of all regarding VPMs and a corresponding SVM. Further, we had also prepared a first draft of a VSB we derived from an already existing model for the VS. Six different stakeholder groups should have been involved: 1) the user/employee requiring a new notebook, 2) the global service desk assisting the user, 3) the team leader approving any orders, 4) the client service team responsible for the procurement and delivery, 5) the external field service executing the delivery and 6) the service managers responsible for this VS. All of our four participants could embody the stakeholder group of the user, whereas the stakeholder group of the external field service was not represented at all. In addition, no participant was working at the global service desk. But, one participant was a team leader, another one part of the client service team and two of them were service managers (see Table 1).

---

[4] www.draeger.com.

## 3.2 Related Work in Artefact Evaluation

To begin with, Prat et al. [21] introduced a taxonomy of evaluation criteria for information systems artefacts in design-oriented research. In this taxonomy, our strategy regarding fits into almost any dimension and evaluation criteria, as we intended a scope as wide as possible for our initial evaluation on *its*VALUE. However, the "generality" evaluation criteria and the entire dimension of "evolution" are not covered at all.

Next, Bork and Roelens [22] introduced a technique for evaluating and improving the semantic transparency of modelling notations. They suggest to split this evaluation process into two phases (evaluating the initial notation and revision) including gathering participant's feedback on the intuitiveness of the notation. Although we did not follow this process, we also asked our participants on the visual elements' intuitiveness.

Further, Buchmann and Karagiannis [23] carried out an evaluation for an ADOxx-based tool as well. Their method aimed at supporting the definition and elicitation of requirements for mobile apps. For their evaluation, they primarily focused on understandability, which we also covered during our study alongside with a perceived usefulness. For both, we asked our participants during the GIs regardingly.

Lastly, Petrusel and Mendling [24] present BPMN-based comprehension evaluation techniques. They used eye-tracking to predict if a model was comprehended or not.

## 3.3 Methodology and Research Questions

According to Döring and Bortz [25], GIs can provide promising information for unexplored and unpredictable outcomes. Thus, we primarily assessed our information and results from GIs. Further, we followed their guidelines to execute them properly: 1) performing a contextual preparation by determining a) the topic, b) the RQs, c) the target group (i.e., the participants and the interviewers) and d) the interview technique (e.g., structured, unstructured, group, single, etc.); 2) defining the GIQs and 3) practicing the GI in advance. The participants' perceptions on the participatory modelling session carried out in advance were explored and addressed during our GIs. Based on the participants' impressions, we investigated if 1) the method fulfils its purpose (see [10]), 2) the notation has a good quality, 3) the ADOxx modeller is a good instance and 4) the modelling session was executed properly.

RQ 2: Does the notation comply with the principles of Moody [17]?
   RQ 2.1: Does it provide a good perceptual discriminability?
   RQ 2.2: Does it provide a good visual expressiveness?
   RQ 2.3: Does it provide a good semantic transparency?
   RQ 2.4: Does it provide a good dual coding?
   RQ 2.5: Does it provide a good graphic economy?

To evaluate our *its*VALUE notation, we derived our RQs from the nine design principles for good notations by Moody [17]. We paraphrased those principles into actual questions. However, the semiotic clarity and cognitive fit are not reasonably evaluable with our case study, as we investigated just a single case instead of multiple ones. Just a single target group cannot provide evidence regarding a suitable notation for many different target groups. Additionally, the complexity management and cognitive integration cannot be reasonably evaluated either, as the participants did not model at all by their own. They just delivered the input and discussed, whereas a host was modelling in parallel. As the participants did not model by their own, they probably could not add any valuable perception on whether the techniques for managing complexity inside and connections across the models are well designed or not. Consequently, we have not considered those four design principles by Moody [17].

RQ 3: Is the ADOxx modeller a good instance of *its*VALUE?
  RQ 3.1: Does ADOxx generally appeal as a fundamental platform?
  RQ 3.2: Are our own procedures supportive and desired?
    RQ 3.2.1: Calculating the cumulative times in a VSB?
    RQ 3.2.2: Receiving the additional SV information for a VSB?
    RQ 3.2.3: Calculating the DR, DA and DU for a process?
    RQ 3.2.4: Detecting waste in a VSB following some patterns?
  RQ 3.3: Is there an intention to use *its*VALUE?
    RQ 3.3.1: Are the participants confident of using it on their own?
    RQ 3.3.2: Can they imagine working with it in the future?
    RQ 3.3.3: Would they use it independently from ADOxx?

As our modeller depends on ADOxx as a fundamental platform, we asked, if ADOxx is perceived as an appealing fundamental platform to model VSs and SV using *its*VALUE. Further, we asked if our own procedures are perceived as valuable contributions to the modelling. Generally, the perceived ease of use cannot be evaluated, as the participants did not work with *its*VALUE or the ADOxx modeller itself. Still, we wanted to check on the perceived usefulness and intention to use by the participants. Lastly, the participants' answers are also highly limited to the participatory modelling session we performed. Therefore, we also questioned the our approach with the following RQs.

RQ 4: Was the modelling session generally executed well?
  RQ 4.1: Was following an entire iteration in a single session appropriate?
  RQ 4.2: Was the separation of work appropriate?
  RQ 4.3: Was the remote execution appropriately performed?
  RQ 4.4: Was the preparation for the modelling session appropriate?

### 3.4 The Guided Interviews (GIs)

As Zucker [20] recommends to design GIQs roughly following the previously defined RQs, we defined our GIQs accordingly. Still, we changed the order of the GIQs compared to the order of the corresponding RQs. Table 2 provides an overview on the order of the GIQs and shows how its blocks are mapped to the corresponding blocks of RQs. Thereby, we aimed at achieving a smoother introduction and farewell for the GIs by placing better suiting questions to those spots recommended by Döring and Bortz [25]. Consequently, we started all GIs with asking how the participants generally perceived the execution of the modelling session. Afterwards, we slightly dived into *its*VALUE by asking how the participants generally perceived the notation. Then, we asked the participants on the *its*VALUE method (see [10]). Lastly, we ended the GIs by asking the participants how they perceived the ADOxx modeller and their intention to use. Generally, we followed the guidance by Döring and Bortz [25] for defining the GIQs. For instance, we avoided expert vocabulary as much as possible when rephrasing our RQs into GIQs to increase the participants' understanding on the concepts we were asking and investigating. Moreover, we also phrased our GIQs as neutral as possible to avoid any framing. To enhance the quality of the participants' answers, we showed them their created models in the ADOxx modeller during the GIs. For the evaluation, we did a deductively structured analysis in the sense of Mayring et al. [26] and considered memory logs (including direct quotes) we created after carefully re-watching recordings of the GIs. As we phrased our RQs to be answered with agreement or disagreement, we just defined three categories for those RQs: *Agreement, Disagreement* and *Neutral*. For the RQs questioning how the participants perceive the different model types of *its*VALUE, we defined the following categories: *Positively, Negatively* and *Neutral*. For instance, if a participant showed a positive response in his wording connected to an aspect covered by our RQs, we assigned a "+" to the regarding spot (see later Tables 3, 4 and 5 in Sect. 4). Vice versa, we did for negative and neutral responses. We just stuck to such a small range of grading, as we wanted to eliminate any bias in different grading levels as much as possible. Any GI was done individually and lasted 20 to 30 min.

**Table 2.** Mapping the RQs and GIQs by blocks.

| GIQ | 1 | 2 | 3 | 4 |
|-----|---|---|---|---|
| RQ  | 4 | 2 | 1 | 3 |

## 3.5 The Pre-test

In advance to the modelling session, we conducted a pre-test followed by a GI. Therefore, we used the exact same environment as in the actual case study. As everything performed well during this pre-test, we stuck to our plan for the real case study. Further, we also tested the GIs with one of the participants from the modelling session pre-test. We encountered difficulties with the interviewer using some expert vocabulary of *its*VALUE when asking the GIQs. We concluded that paraphrasing such terms would increase the understandability of the GIs. Thus, we slightly adapted our GIs and conducted a second pre-test with a person completely unrelated to any technical topics to check the understandability of our GIs. To provide the new interviewed person with the capability to answer the GIQs properly, we held an introductory presentation on *its*VALUE lasting 30 min in advance by explicitly demonstrating our ADOxx Modeller. This new person felt comfortable with answering any GIQ and did not struggle with any vocabulary.

## 4   Evaluation Results

In the following subsections, we present our results we derived from our GIs as described in Sect. 3.4. We start with the *its*VALUE notation, followed by the ADOxx Modeller and end with the execution of our modelling session.

### 4.1   Results on the *its*VALUE Notation

Table 3 shows how the participants' answers from the GIs correspond to the RQs of block 2. These results reveal a big difference between the perceptions of the participants already familiar with modelling (A, B and D) and the one that is not (C). A, B and D perceived the notation as generally good, the visual differences as easy to perceive and liked the wide range of different visualizations. However, B and C stated that the colours of some objects may look too similar (e.g., information-related processes and IT-Systems). Further, A perceived that the relations are harder to differentiate than objects. Moreover, A, B and D perceived the notation across all models as intuitive. They liked especially the colours for the RF, CPL, SS and HS, as they remind of traffic lights which supports understanding the meaning those KPIs embody. Although A states that the VSB is very well structured, C perceives it and the SVM as hard to understand at all. Generally, C also perceived the range of different visualizations as too complex. However, C enjoyed the VPM (especially its graphic economy) and agreed with all other participants that the amount of text is well balanced

**Table 3.** Results derived from the GIs on the notation of *its*VALUE.

| RQ | P A | P B | P C | P D | $\sum +$ | $\sum n$ | $\sum -$ |
|---|---|---|---|---|---|---|---|
| 2.1 | + | + | n | + | 3 | 1 | 0 |
| 2.2 | + | + | − | + | 3 | 0 | 1 |
| 2.3 | + | + | − | + | 3 | 0 | 1 |
| 2.4 | + | + | + | + | 4 | 0 | 0 |
| 2.5 | + | + | − | + | 3 | 0 | 1 |

**Key:** "P" = Participant; "+" = Agreement ∨ Positively;
"n" = Neutral; "−" = Disagreement ∨ Negatively

and not too much or too less across all models. Further, all participants perceive the VSB as overwhelmingly (at least in the beginning). However, A, B and D perceive it as okay, manageable and even required, whereas C disagrees on that. Moreover, C disliked the amount of objects inside the SVM. To conclude, there are differences in the perception of the notation for the different model types, although we designed it combined. Thus, we considered the overall participants' perception for Table 3. To conclude, for three out of four participants, the *its*VALUE notation seems to comply with (at least) five principles by Moody [17].

## 4.2   Results on the *its*VALUE ADOxx Modeller

Table 4 shows how the answers of the participants from the GIs correspond to the RQs of block 3. It reveals that all participants perceive the ADOxx modeller as supportive. However, A questioned its mechanics for designing a new VS (e.g., missing recommendation functionalities). Further, A dislikes ADOxx as a fundamental platform, as A questioned its potential integration to already common systems. Oppositely, B, C and D perceive ADOxx as well and useful. However, B limits his impressions due to having no experiences with using it on their own. Further, all participants perceived all procedures as useful and supportive. For the *"Cumulative Time Calculator"*, they underline its importance and usefulness e for decreasing time consumption in a VS on the fly. For the *"Receive Value Information"*, they state that it enriches the VS further with very supportive and useful information. Additionally, they perceive the *"Data Information Calculator"* as good and useful. However, A desires an extended calculation for the entire VS instead of just single processes, whereas B limits

**Table 4.** Results derived from the GIs on the ADOxx modeller.

| RQ | P A | P B | P C | P D | $\sum+$ | $\sum n$ | $\sum-$ |
|------|-----|-----|-----|-----|-----|-----|-----|
| 3.1 | − | n | + | + | 2 | 1 | 1 |
| 3.2 | + | + | + | + | 4 | 0 | 0 |
| 3.2.1 | + | + | + | + | 4 | 0 | 0 |
| 3.2.2 | + | + | + | + | 4 | 0 | 0 |
| 3.2.3 | + | n | + | + | 3 | 1 | 0 |
| 3.2.4 | + | + | + | + | 4 | 0 | 0 |
| 3.3 | + | + | + | + | 4 | 0 | 0 |
| 3.3.1 | − | − | − | − | 0 | 0 | 4 |
| 3.3.2 | + | + | + | + | 4 | 0 | 0 |
| 3.3.3 | + | + | + | + | 4 | 0 | 0 |

**Key:** *"P" = Participant; "+" = Agreement ∨ Positively;*
*"n" = Neutral; "−" = Disagreement ∨ Negatively*

**Table 5.** Results derived from the GIs on the execution of the modelling session.

| RQ | P A | P B | P C | P D | $\sum +$ | $\sum n$ | $\sum -$ |
|---|---|---|---|---|---|---|---|
| 4.1 | + | $n$ | + | $n$ | 2 | 2 | 0 |
| 4.2 | + | + | + | + | 4 | 0 | 0 |
| 4.3 | + | + | + | + | 4 | 0 | 0 |
| 4.4 | $n$ | + | $n$ | + | 2 | 2 | 0 |

**Key:** "P" = Participant; "+" = Agreement ∨ Positively;
"n" = Neutral; "−" = Disagreement ∨ Negatively

his answer due to not feeling capable of properly evaluating it. Lastly, all participants perceive the *"Waste Detector"* as outstandingly supportive. Generally, C desires some kind of additional *"Cost Calculator"* that can visualize cost savings as a result of dealing with waste. Generally, all participants perceived *its*VALUE as very suitable for future applications (for both designing new and improving existing VSs). Whereas D can imagine introducing and using it instantly in the company, A makes it dependent on the maintenance *its*VALUE may receive in the future. Although no participant feels capable of using *its*VALUE on their own, they all can imagine and even desire a workshop to achieve that. Further, they all want to work with *its*VALUE again. B can even imagine of becoming a modelling host guiding a similar modelling session. Lastly, all participants would also use *its*VALUE independently from ADOxx, as long as there was an alternative. Without using a software (e.g., using a plastic wall and coloured pieces of paper), B only perceives the VPM and SVM as applicable. According to B, the VSB requires a software tool especially due to the calculated times. To conclude, all four participants seem to perceive the ADOxx modeller as a good instance of *its*VALUE.

### 4.3    Results on Our Modelling Session Execution

Table 5 shows how the participants' answers from the GIs correspond to the RQs of block 4. It shows a satisfaction by the participants with the execution of and guidance through the modelling session. Although A demanded a more strict execution of the procedure without jumping between different types of tasks, B stated that this was not a problem at all due to the good guidance. Further, A stated that the moderation dived in too deep into some mechanics of the ADOxx modeller at some points. A and C perceived the duration of 2:30 h for this session and an entire iteration as perfect, whereas B and D consider it as too short but still okay due to the good preparations in advance. Generally, D perceived the objective as "tangible". Moreover, the separation of tasks and remote execution via Microsoft Teams was positively perceived by all participants. As they all have already known each other, they all perceived their collaboration as good and productive. B considers a similar approach even as reasonable for potential further applications. Moreover, all participants can imagine that a similar procedure could also (or even better) work in present. Further, B and D perceived the preparations for the modelling session as well (the introduction to the topic) or even "impressing" (the prepared models to achieve a smoother start). To conclude, all participants perceived the execution of the modelling session as good.

## 5   Discussion

We wanted to check if the participants (dis-)like something because of a) how it actually is or b) just the way we designed the modelling session (i.e., the contextual condition). As they often mentioned that something was still good due to the detailed guidance and explanations, we believe that the modelling session was well executed. Thus, the expressiveness of the answers given may also have a good quality.

However, the participants' perceptions are limited as none of them has used or even known about ADOxx before. Thus, they cannot provide that expressive input to answer if ADOxx is an appealing fundamental platform or not. Still, they all encountered the software of ADOxx and our modeller extensively just from an observational point of view. Thus, their perceptions may still be valuable. We believe that the ADOxx modeller (as an instance of *its*VALUE) had an impact on how the participants perceived the method and notation, as it directly implements what *its*VALUE is all about. For instance, they were impressed by the possibility ADOxx offers through our procedures and dynamic scripts we implemented to support the users. As all participants have an intention to use *its*VALUE and especially the ADOxx modeller, we believe it is a good instance for evaluating *its*VALUE the method and notation.

With regard to the answers given by participant C, we believe that the *its*VALUE notation (especially for the VSB) had a negative influence on how C perceived the method of *its*VALUE for understanding and defining VSs. As C perceived the notation for the VSB as too complex and incomprehensible, C concluded that understanding and defining a CVS is hard to achieve using *its*VALUE overall. Oppositely, this was not the case for the other three participants. In contrast to C, they all claim to be familiar with modelling. Thus, the *its*VALUE notation (especially for the VSB) may only suit those already familiar with similarly modelling techniques (e.g., BPMN, 4EM, etc.). Further, the quality of the *its*VALUE notation seems to differ for its different model types. As the VSB contains much more different concepts than the VPM or SVM, especially its complexity management and cognitive fit are probably harder to handle and to accomplish. We believe, that the notation for the VSB even underperforms compared to both the VPM and SVM. Unfortunately, we have not explicitly asked for the perceived differences of the notations for all three model types due to time limitations.

Considering our previous results from [10] on the *its*VALUE method (indicating that the method matches its hypothesis in supporting understanding and enhancing VSs for all participants), the *its*VALUE notation and ADOxx modeller seem to fit their overall requirements. Following the AMME methodology by Buchmann and Karagiannis [27, 28] (an iterative process for method engineering), we defined new modelling requirements specifically scoped at enhancing the VSB. Our results revealed that the VSB easily tends to become too "full" (i.e., it may contain too many different objects). This may decrease the overview on the actual flow of a VS. In turn, enhancing and especially understanding VSs may suffer. To address this issue, some kind of *Process Specification Model (PSM)* could be introduced sourcing out many objects. Following this idea, any process part of a VS would receive its own PSM. Similarly to the VPM, the process itself would embody the core of that model. Any other object the process somehow interacts or interfaces with would be placed around that process and linked to it. Such information visually stored inside the PSMs could be transferred to the VSB (like the VPMs). There, it would be stored inside internal annotation tables of the processes (accessible via double click in ADOxx). This would make the VSB leaner, as it would just present the actual flow between all processes, and no information would be lost due to our dynamic script.

## 6   Limitations

Our case study and in turn evaluation on *its*VALUE are limited in several aspects. Firstly, we have just investigated one single case. Consequently, a generalization from that particular single case to several other different ones is not reasonably possible according to Zucker [20] and Gerring [29]. To reasonably draw a generalization, many additional case studies must be performed in several scenarios. Further, we have just gathered findings for *its*VALUE on the perceived efficacy and intention to use, whereas information on the actual efficacy and usage still remain unknown. This must be also investigated in future work. Moreover, we have just investigated one of two possible paths of the *its*VALUE method and performed just one single iteration that even did not cover all stakeholders and spots of the investigated VS. In consequence, our findings just deliver a first implication and do not provide any reasonable evidence, especially for the suitability of *its*VALUE for designing an entirely new service and VS. In addition, the notation was not explicitly investigated for each model type of *its*VALUE separately. However, their quality seems to differ according to the answers of the participants. Thus, their statements regarding the notations for each model type are mainly limited to their general overall perception. In future work, each model type should be investigated separately regarding its notation quality. Additionally, as the participants did not actually use the ADOxx modeller by their own, their perception and thus our results on it are also limited. Moreover, the available time for both the modelling session and the GIs was limited and in turn short. As some participants suggested that more time would be required to cover all stakeholders and aspects of a VS (which we actually did not achieve), this should be carefully considered in future work (especially when performing several iterations of *its*VALUE). Further, we have not considered the desirability bias during the GIs where participants may have the intention to not insult the interviewer, especially in face-to-face communication. But as we did everything via Microsoft Teams, this may have mediated the influence of the desirability bias. Lastly, we want to mention that we could just encourage four people to participate and in turn to deliver input. This limits the expressiveness of our results. To conclude, we understand our findings just as an initial direction that seems to promising to follow.

## 7  Conclusion and Outlook

*its*VALUE in combination with the ADOxx modeller supports comprehensive modelling and analysis of IT-Service related to VSs. Based on its consideration of ITIL 4 concepts, it may support practitioners in adopting that standard. Our first case study revealed a great relevance of the *its*VALUE models and analysis results for value-oriented IT-Service modelling according to the participants involved. Further, they showed interest in future use of the *its*VALUE method and ADOxx modeller. Moreover, they perceived all three *its*VALUE model types as good and supportive (with the exception of one participant perceiving the VSB as too complex and incomprehensible). Considering the answers of all participants, this may reveal that the *its*VALUE notation suits experienced modellers better than inexperienced. Additionally, they perceived the ADOxx modeller as powerful and beneficial. Lastly, all participants were satisfied with the way we executed the participatory modelling session.

Considering the limitations of the discussed case study, further evaluations and iterations of method engineering are required as suggested by Buchmann and Karagiannis [27,28]. For instance, the introduction of a PSM could enhance the *its*VALUE method, notation and ADOxx modeller. However, future work must explore if such changes are desired and even beneficial. Firstly, such a PSM should be carefully developed conceptually. In turn, its influence on the *its*VALUE method framework must also be explored and lastly be added to the ADOxx modeller. Moreover, future work should also keep in mind the questions by Lewin et al. [30]: *1) Which design principles should be developed?, 2) How does a waste free industry 4.0 actually looks like?* and *3) Which systems or logic actually add value?.* So far, we have just designed a method and notation We implemented both with a corresponding ADOxx modeller. These artefacts enable applicants explicitly considering SV when designing or improving VSs from a more generic point of view. Hence, the nature of *"waste"* and *"value"* should also be investigated.

Future evaluations will help to better adjust and refine the approach for practitioners. Having the modeller freely available at OMiLAB[5] assures access for and involvement of potential users. An important next step will be improving complexity handling to make *its*VALUE suit complex scenarios better. Further, developing method guidelines fitting the needs of practitioners still remains a task to accomplish.

---

[5] www.omilab.org/activities/omilab-book-series/volume2/details/?id=42.

# References

1. Axelos: ITIL 4 Strategic Leader: Digital and IT Strategy (PDF). ITIL Managing Professional Series. The Stationery Office (2020)
2. Axelos: ITIL 4 Create, Deliver and Support. ITIL 4 Managing Professional Series. The Stationery Office (2020)
3. Axelos: ITIL 4 Managing Professional Drive Stakeholder Value. ITIL 4 Managing Professional Series. The Stationery Office (2020)
4. Axelos: ITIL 4 Direct, Plan and Improve. ITIL 4 Managing Professional Series. The Stationery Office (2020)
5. Axelos: ITIL 4 High-Velocity IT. ITIL 4 Managing Professional Series. The Stationery Office (2020)
6. Richter, H., Lantow, B.: IT-service value modeling: a systematic literature analysis. In: Abramowicz, W., Auer, S., Stróżyna, M. (eds.) BIS 2021. LNBIP, vol. 444, pp. 267–278. Springer, Cham (2022). https://doi.org/10.1007/978-3-031-04216-4_24
7. Schoenwaelder, M., Szilagyi, T., Baer, F., Lantow, B., Sandkuhl, K.: It self-service blueprinting a visual notation for designing it self-services. In: Bork, D., Lantow, B., Grabis, J (eds.) CEUR Workshop Proceedings, vol. 2238, pp. 88–99. CEUR-WS (2018)
8. Hartmann, L., Meudt, T., Seifermann, S., Metternich, J.: Value stream method 4.0: Holistic method to analyse and design value streams in the digital age. Procedia CIRP **78**, 249–254 (2018)
9. Hartmann, L., Meudt, T., Seifermann, S., Metternich, J.: Value stream design 4.0: designing lean value streams in times of digitalization and industrie 4.0 [wertstromdesign 4.0: Gestaltung schlanker wertstroeme im zeitalter von digitalisierung und industrie 4.0]. ZWF Zeitschrift fuer Wirtschaftlichen Fabrikbetrieb **113**(6), 393–397 (2018)
10. Richter, H., Lantow, B.: *its*VALUE - a method supporting value-oriented ITSM. In: Buchmann, R.A., Polini, A., Johansson, B., Karagiannis, D. (eds.) BIR 2021. LNBIP, vol. 430, pp. 133–149. Springer, Cham (2021). https://doi.org/10.1007/978-3-030-87205-2_9
11. Richter, H.D., Lantow, B., Pröpper, T.: itsVALUE: modelling and analysing value streams for it services. In: Karagiannis, D., Lee, M., Hinkelmann, K., Utz, W. (eds.) Domain-Specific Conceptual Modeling, pp. 161–183. Springer, Cham (2022). https://doi.org/10.1007/978-3-030-93547-4_8
12. Meudt, T., Leipoldt, C., Metternich, J.: Der neue blick auf verschwendungen im kontext von industrie 4.0: Detaillierte analyse von verschwendungen in informationslogistikprozessen. Zeitschrift für wirtschaftlichen Fabrikbetrieb **111**(11), 754–758 (2016)
13. Meudt, T., Metternich, J., Abele, E.: Value stream mapping 4.0: holistic examination of value stream and information logistics in production. CIRP Ann. **66**(1), 413–416 (2017)
14. Heger, S., Valett, L., Thim, H., Schröder, J., Gimpel, H.: Value stream model and notation-digitale transformation von wertströmen. In: Wirtschaftsinformatik (Zentrale Tracks), pp. 710–724 (2020)
15. Sandkuhl, K., Wißotzki, M., Stirna, J.: Begriffe im umfeld der unternehmensmodellierung. In: Sandkuhl, K., Wißotzki, M., Stirna, J. (eds.) Unternehmensmodellierung: Grundlagen, Methode und Praktiken, pp. 25–40. Springer, Heidelberg (2013). https://doi.org/10.1007/978-3-642-31093-5

16. Geist, M.R.: Using the delphi method to engage stakeholders: a comparison of two studies. Eval. Program Plann. **33**(2), 147–154 (2010)
17. Moody, D.: The "physics" of notations: toward a scientific basis for constructing visual notations in software engineering. IEEE Trans. Software Eng. **35**(6), 756–779 (2009)
18. Karagiannis, D., Kühn, H.: Metamodelling platforms. In: EC-Web, vol. 2455, p. 182 (2002)
19. Baxter, P., Jack, S., et al.: Qualitative case study methodology: study design and implementation for novice researchers. Qual. Rep. **13**(4), 544–559 (2008)
20. Zucker, D.M.: How to do case study research. School of Nursing Faculty Publication Series, p. 2 (2009)
21. Prat, N., Comyn-Wattiau, I., Akoka, J.: Artifact evaluation in information systems design-science research-a holistic view. In: PACIS 2014 Proceedings, no. 23 (2014)
22. Bork, D., Roelens, B.: A technique for evaluating and improving the semantic transparency of modeling language notations. Softw. Syst. Model. **20**(4), 939–963 (2021)
23. Buchmann, R.A., Karagiannis, D.: Modelling mobile app requirements for semantic traceability. Requirements Eng. **22**(1), 41–75 (2017)
24. Petrusel, R., Mendling, J.: Eye-tracking the factors of process model comprehension tasks. In: Salinesi, C., Norrie, M.C., Pastor, Ó. (eds.) CAiSE 2013. LNCS, vol. 7908, pp. 224–239. Springer, Heidelberg (2013). https://doi.org/10.1007/978-3-642-38709-8_15
25. Döring, N., Bortz, J.: Forschungsmethoden und Evaluation. Springerverlag, Wiesbaden (2016)
26. Mayring, P., Fenzl, T.: Qualitative Inhaltsanalyse. In: Baur, N., Blasius, J. (eds.) Handbuch Methoden der empirischen Sozialforschung, pp. 633–648. Springer, Wiesbaden (2019). https://doi.org/10.1007/978-3-658-21308-4_42
27. Buchmann, R.A., Karagiannis, D.: Agile modelling method engineering: lessons learned in the ComVantage research project. In: Ralyté, J., España, S., Pastor, Ó. (eds.) PoEM 2015. LNBIP, vol. 235, pp. 356–373. Springer, Cham (2015). https://doi.org/10.1007/978-3-319-25897-3_23
28. Karagiannis, D., Burzynski, P., Utz, W., Buchmann, R.A.: A metamodeling approach to support the engineering of modeling method requirements. In: 2019 IEEE 27th International Requirements Engineering Conference (RE), pp. 199–210 (2019)
29. Gerring, J.: What is a case study and what is it good for? Am. Political Sci. Rev. **98**, 341–354 (2004)
30. Lewin, M., Voigtlander, S., Fay, A.: Method for process modelling and analysis with regard to the requirements of industry 4.0: an extension of the value stream method. In: Proceedings IECON 2017–43rd Annual Conference of the IEEE Industrial Electronics Society, 2017-January, pp. 3957–3962 (2017)

# Retrospective Considerations on Data Flow and Actor Modeling in Business Process Diagrams

Marite Kirikova(✉) ⑩

Department of Artificial Intelligence and Systems Engineering, Riga Technical University, Riga, Latvia
Marite.Kirikova@rtu.lv

**Abstract.** The BPMN 2.0 for modeling collaborative business processes is already a decade old, and it has gained considerable attention from researchers and practitioners with respect to its benefits and drawbacks. The impossibility of modeling data flows properly (input and output data combinations of activities shown only as comments and data associations used instead of data flows) as well as problems with reflecting activities which are executed by several actors, are among the often-mentioned limitations of the BPMN 2.0. Some business process modeling languages and tools that existed before the BPMN, however, did provide solutions to the above-mentioned limitations. It might be beneficial to consider these solutions in the development of the next versions or dedicated extensions of the BPMN.

**Keywords:** Business process modeling · BPMN · IDEF0 · IBM WebSphere · GRAPES BPM

## 1 Introduction

This empirical paper concerns data flow and actor modeling in business process models. Representation of data flows in business process models helps us to see how data created by one activity is used in other activities and to ensure that all data necessary for producing the output of an activity is available as inputs(s) of this activity. Activities of a business process can be performed either by human or artificial actors (e.g., software, robots, etc.). The availability of actors, in the model, gives an opportunity to automatically rearrange the activities so that the division between the activities performed by different actors can be analyzed and assessed. These aspects are especially important when business processes are used as the basis for requirements derivation for information technology solutions [1]. While the most popular notation of business process modeling, BPMN 2.0, gives only visual means for data and actor representation and for relating activities to actors [2], there are a number of approaches that have been proposed for more comprehensive data modeling, for instance, [3]; also there are several business process modeling tools capable of relating business activities to data models and actors;

Ē. Nazaruka et al. (Eds.): BIR 2022, LNBIP 462, pp. 86–95, 2022.
https://doi.org/10.1007/978-3-031-16947-2_6

for instance, tools from ADO [4] and from No Magic [5] families. The fact that the tools have these capabilities proves pragmatically that it is necessary to handle data and actors as "first class citizens" in the business process model rather than just having the possibility of representing them in the model. However, the way of handling data and actors varies between the tools. In most cases, data handling is limited just to pointing to the data objects as inputs and outputs of specific activities, neglecting the visualization of the data flow. Regarding actors, usually they can be addressed by relating them to organizational models via swimlanes, or they can be represented next to the activities.

The research approach taken in this paper is "learning from the past". Some of the issues that are not sufficiently addressed in modern notations and tools were comfortably solved in older business process modeling approaches. In this paper we will reflect on these solutions from the past with the purpose of seeing whether it would be possible to incorporate them in today's notations and tools.

The paper is organized as follows. The related work regarding the "retro" approaches is discussed in Sect. 2. The approach embracing some retro features for data flow and actor modeling in BPMN-like notation is proposed in Sect. 3. The proposed approach is evaluated in Sect. 4. Brief conclusions are presented in Sect. 5.

## 2  "Retro" Approaches Considered

In order to learn from past experiences, the modeling approaches were chosen based on two criteria (1) these had to be modeling languages (notations, tools) that existed before BPMN and (2) the languages (notations, tools) had to have the means for data flow modeling and actor representation. Three approaches have been chosen: IDEF0 that belongs to the IDEF modeling language group and has been used for modeling technological processes, IBM WebSphere approach, and GRAPES BPM 4.0, which is the language that grew out of the formerly most popular language EPC (event process chains) [6]. The descriptions of IDEF0 [7], WebSphere approach [8] and GRAPES BPM 4.0 [9] are freely available.

### 2.1  Data Flows and Actors in IDEF0

The IDEF0 diagram is a well-known and well-specified process representation language with a strict specification [7]. It used to be the one of the dominant notations in business process modeling up to the end of the previous century; afterwards giving the floor to languages with less strict rules on activity sequence representation. The IDEF0 diagram allows the representation of both the data flows and the actors. It distinguishes between data flows and control flows by the location of the point linking a flow to an activity. Examples of IDEF0 diagrams used for business process modeling are available in [10]. One of the main drawbacks of this notation is its very dense network of relationships (linkages) of data and control flows that makes it difficult to comprehend the processes if they are represented by the non-artificial number of activities [11]. It is important that the IDEF0 diagrams can show an arbitrary number of actors for each activity. The actors can be shown with arrows going bottom up to the activities. This form of representation fits well with the original diagram layout of IDEF0 but becomes cluttered for other layouts.

## 2.2   Data Flows and Actors in IBM WebSphere Business Modeler

Regarding data flows and actors, IBM WebSphere is considered as a tool. An example of data flow and actor representation in IBM WebSphere Business Modeler 6.0.1 is available in [12]. The data flow is represented in-line with the control flow; and it can be attached to the branching (triggering) elements. The actors can be attached to the process model by relating them to the activities. The actors can be represented beneath the activities, for instance. A specific, and important for this paper, feature of the WebSphere tool is its capability to reconstruct the business process model according to specific attributes (either actors or other business items). This feature enables multi-perspective [13] or multi-dimensional [1] business process analysis that is essential for seeing business process compliance with different context factors or attributes. Thus, once actors are related to activities, the tool can generate lanes wherein the actors' activities are represented. This capability to reorganize process representation according to chosen business items is very rare in business process modeling tools; however, it provides an opportunity to view the process from different perspectives and therefore is very useful.

## 2.3   Data Flows and Actors in GRAPES BPM and GRADE Tool

The GRAPES BPM language and the GRADE tool, which supported it, currently seems to be out of use as only a few traces of them are seen on the Web [9]. However, this language used to be well regarded by practitioners and was taught at several universities. GRAPES BPM showed data flows in-line with control flows allowing them to be related to triggering elements. It is important that, in this language, the triggering element was located inside the activity box thus not occupying an extra space. The triggering condition was reflected in the form of a logical expression, thus enabling complex variants of flow combinations for the triggering to take place. A branching mechanism was located outside the activity box. An individual actor or a group of actors could be assigned to each activity. Similar, to WebSphere, a swimlane representation of a process could be generated by having a separate lane for each unique actor and each unique combination of actors assigned to activities. All activities without actors were allocated to a separate lane signaling that there is nobody assigned to the work to be done.

## 2.4   Comparing the "Retro" Approaches to BPMN 2.0

Business process modeling languages and notations have been compared by several researchers using different criteria. For instance, in [14], BPMN 2.0 is compared to three notations for manufacturing processes: IDEF3, PSL, and VSM. One of the comparison criteria used is "information flows". The authors admit that, to show these flows, IDEF3 needs IDEF0 and IDEF1 diagrams, PSL needs extensions, VSM can show the flows explicitly without their internal structure, and BPMN 2.0 can show the flows explicitly, relying on extensions for internal structure. Here we should clarify that there are two ways in which the information (data) flow can be shown in a BPMN model – as a message flow between pools or as data objects associated with activities inside a pool. The authors of [14], obviously, refer to message flows in their work.

In [15], BPMN is compared to EPC, UML activity diagrams, RAD, and IDEF by such criteria as expressiveness, readability, usability, user friendliness, formality, versatility, tools support, flexibility, concision, ease of learning, possibility to induce innovation, possibility to evolve, and applicability for collaborative work. These criteria are derived from seven other comparisons of business process modeling languages that are referred to in [15]. The criteria in [15] do not directly address the data flows and actors, however, they will be used in this paper when evaluating the proposed approach. In [16], 12 approaches are compared based on Bunge, Wand, and Weber's information systems ontology. This comparison also does not directly address data flows and actors.

From the perspective of this paper, the "retro" approaches are compared to BPMN 2.0 in Table 1.

**Table 1.** Comparison of "retro" approaches and BPMN 2.0.

| | IDEF0 | IBM WebSphere BM | GRAPES BPM | BPMN 2.0 |
|---|---|---|---|---|
| Representation of data flow | Can be represented *directly*, as a flow also *showing outputs/inputs* | Can be represented by relating business items to activities | Can be represented *directly* via the links | As message flows (directly); and as data objects associated to activities |
| Data and triggering conditions | Not explicitly shown (only as the shape of a data flow). *Distinguishes from control flows* by placement position | *Data flows can be related to triggering representation elements.* The triggering elements are represented outside the activity boxes | *Data flows can be related to triggering elements, triggering elements are inside the activity box represented as logical expressions* | Data is not related to triggering elements which are situated outside the activity elements. Data combinations can be explained as comments only, distinguishes data from control flow |
| Representation of actors | At the bottom of the diagram with bottom-up arrows. Arbitrary number of Actors per activity can be represented | Possible to represent below the activity box | *Actors can be represented inside the activity box* | Only as swimlanes |
| Swimlane view generation for actors | Not applicable | Possible | *Possible for each unique actor or group of actors* | Not applicable |

The following criteria are used for comparison: representation of data flow, data and triggering conditions, representation of actors, and swimlane view generation for actors. IDEF0, IBM WebSphere Business Modeler (BM), and GRAPES BPM (and GRADE tool) are compared.

When looking at the differences between BPMN 2.0 and other approaches we can see several proven possible ways of overcoming deficiencies in modeling of data flows and actors. These possibilities are as follows:

- Possibility to reflect data flows directly
- Possibility to have triggers for data flows
- Possibility to reflect actors together with the activities
- Possibility to automatically generate swimlane views if actors are assigned.

In the next section these possibilities will be illustrated graphically. The "retro" features further discussed in Sect. 3 are highlighted in italic in Table 1.

## 3  Embracing "Retro" Features

To illustrate the possibilities of data and actor modeling discussed in the previous section, a small fragment of a business process model made in BPMN by a business process modeling course student will be used. This fragment, reflected in Fig. 1, has some features which are not supported by BPMN 2.0 but are available in ADONIS [4], the business process modeling tool which was used for creating this fragment.

These additional features are a possibility to show RACI (responsible, accountable, consulted, informed) matrix elements below the activity and relating inputs and outputs to the model (the arrows attached to the right upper corner of activity boxes). The input and output allocation feature of the tool is out of the scope of this paper. The fragment shows four activities and three associated data objects.

The information available in Fig. 1 is further reflected in the Fig. 2, which is built using "retro" features discussed in the previous section and reflects the proposed approach for business process modeling.

In Fig. 2, the actors are included in the activity box as it is done in GRAPES BPM and GRADE tool. Only the actors with role "responsible" (R) are shown in Fig. 2. In this way it is possible to see the actor of the elements regardless of pools and lanes. Another benefit is that it is also possible to allocate several actors to one activity, and even to do it using logical expressions (not shown in Fig. 2) as it was possible in the GRAPES BPM language. Data flows are used instead of associations to data objects, saving space and making the diagram less cluttered. Data triggering conditions are introduced. In this way there is no need to have two parallel representations (one for inputs and outputs, another for data objects) as was the case in Fig. 1. The proposal is to keep the data and sequence flows separate, which would differ from the "retro" methods. Nevertheless, the triggering conditions for data flows are available and are organized similarly as in GRAPES BPM, i.e., inside the activity box. To have an easy and clear description of triggering logic, the data flow identifiers are introduced and these provide automatically assigned labels that are used in logical triggering expressions.

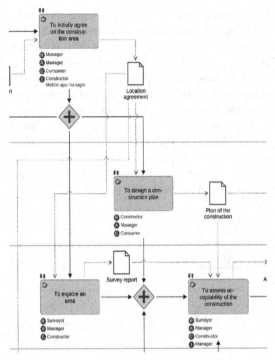

**Fig. 1.** Example: the fragment from the student's created business process model in BPMN (using the business process modeling tool ADONIS [4]).

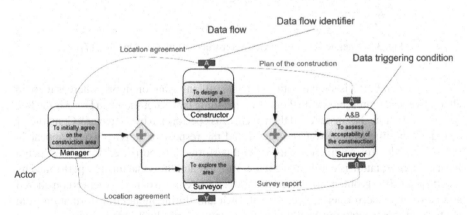

**Fig. 2.** Enhanced BPMN2.0 model fragment (partly created with Bee-Up tool [17]).

The information reflected in Fig. 2 is sufficient for automatically generating the view that is similar to the one shown in Fig. 1. While the representation in Fig. 2 is richer information-wise (the data triggering conditions are available), it occupies less space and is less cluttered. In modeling tools, it also could be possible to easily switch between sequence flows only, sequence flow and data flow representations (as in Fig. 2), and sequence flow and data object representations as currently in BPMN 2.0 (Fig. 1).

One more capability of "retro" tools that brings benefits in business process analysis is the possibility of generating swimlane views from the swimlane-free models. While the IBM WebSphere business process modeling tool allows such generation for any attributes (business items) assigned to activities, the GRADE tool supporting the GRAPES BPM language provides this functionality only for actors. The view that could be generated from information available in Fig. 2 is shown in Fig. 3. Here the three actors (roles) Constructor, Surveyor, and Manager, shown in Fig. 2, are represented as the lanes in the pool. The sequence of activities in Fig. 3 is the same as in Fig. 2, however, the outlook differs, which helps us to see the process from a different perspective. Figure 2 provides functional perspective and Fig. 3 is oriented on actors [1, 13].

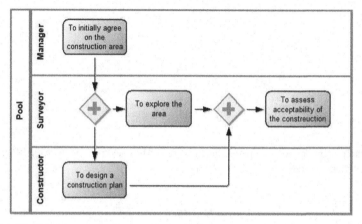

**Fig. 3.** Generated swimlane view (constructed with Bee-Up tool [17]).

The view in Fig. 3 is shown with "switched off" information flows, whereas it could alternatively be represented with them. Another thing that is not shown in Fig. 3 is the line with several actors. For instance, if there were a task performed by Manager&Constructor in Fig. 2, the line Manager&Constructor and the respective activity would appear in Fig. 3. The idea for swimlanes with several performers is borrowed from [9]. Such a form of representation is comfortable if there are few activities that must be performed by several actors (the problem not solved in BPMN 2.0). A group of performers is considered as a new unique performer and receives its own lane during model transformation from the form of representation illustrated in Fig. 2 to the form illustrated in Fig. 3.

The proposed approach allows for the relating of elements of a business process model to other models representing the context of the business process, such as orga-nizational structure and data or document models. Discussion on the necessity of such relationships has been published in several scientific papers (for instance, [3, 18, 19]); and means for reflecting these relationships have been implemented in some enterprise and business process modeling tools, e.g., ADONIS [4]. However, an extended discussion on reflecting a business process model in its context is beyond the scope of this paper.

## 4 Evaluation of the Proposed Approach

The approach proposed in the previous section is evaluated using criteria from [15], namely, we discuss expressiveness, readability, usability, user friendliness, formality, universality, tools support, flexibility, concision, ease of learning, possibility to induce innovation, possibility to evolve, and applicability for collaborative work of the proposed approach. Logical argumentation is used as a means for evaluation. This is a preliminary evaluation, because the practical implementation of the tool is necessary in order to obtain evidence from practice about the applicability of the approach.

The approach is more *expressive* than the native BPMN 2.0, because it reflects more information (data flow, data triggering conditions, actors). The approach can be positively evaluated, also, from the point of view of *readability*, because (1) additional information and (2) the possibility of switching between different views can make it easier for people to understand and interpret the model. Both of the above-mentioned features also contribute to *usability* as they help with the handling of complexity in modeling. These features might also contribute to *user friendliness*, however user friendliness is a rather subjective criterion, because evaluation regarding pleasantness and intuitiveness that characterize it depends on the former experiences of users and may vary between different user groups. The proposed approach contributes to *formality* by clear definition of data flows and their triggering conditions. The proposed approach cannot be evaluated on *universality*, as it is not yet in use. Regarding *tool support*, it may be said that the capabilities of former and current enterprise modeling and business process modeling tools allow for tools support for the proposed approach. The approach greatly contributes to *flexibility* because it can be used in different scenarios and represent distinct features. The approach helps with *concision* as it can show various facets of a business process with a smaller set of elements (data flows instead of data objects and two association links). *Ease of learning* of the approach cannot yet be evaluated as it depends on previous experiences of learners and experiments are needed to prove the ease of learning. Hopefully, the approach will serve as an *innovation inducer* because it embraces some features from the business process modeling approaches used in the previous century, which might foster the arising of a new wave in business process modeling approaches, and tools that are more capable of data and information flow representation. The approach is *evolutionary*, because it gives an opportunity to work on optimal representations when generating swimlane views and for the representation of triggering conditions. There are no specific benefits of this approach for *collaborative work*, however, neither does it have any disadvantages regarding this evaluation criterion.

Thus, the preliminary evaluation of the proposed approach shows that it might be useful in business process modeling and solve some of the problems faced with BPMN 2.0 regarding data flow and business process actor modeling.

# 5 Conclusions

This paper provides a retrospective view on business process modeling languages, notations, and tools and proposes that some "retro" features be integrated with BPMN 2.0. The proposed approach gives an opportunity to model data or information flows (also with triggering conditions) and actors. It also allows for generating swimlane views from the models, with separate swimlanes, not only for specific actors, but also for groups of actors; thus, solving the problem (common in current swimlane representations) regarding the representation of activities that refer to several swimlanes.

The approach in this paper is described solely from the business process model perspective and does not include discussion on relating data flows to data and document models, or relating actors to organizational models. Neither it discusses the relationships between data flows and entity life cycles. These discussions are purposely left out of the scope of this paper which focuses purely on data flows and actors inside the business process model. Further research is intended regarding distinguishing between data, information, and knowledge flows in business process models [20] and relationships between business process models and information demand analysis in information logistics [21].

# References

1. Businska, L., Kirikova, M.: Multidimensional modeling and analysis of business processes, In: Advances in Databases and Information Systems: 13th East-European Conference (ADBIS 2009): Associated Workshops and Doctoral Consortium: Local Proceedings, Latvia, Riga, RTU, pp. 33–47 (2009)
2. About the Business Process Model and Notation Specification Version 2.0. https://www.omg.org/spec/BPMN/2.0/
3. Combi, C., Oliboni, B., Weske, M., Zerbato, F.: Conceptual modeling of inter-dependencies between processes and data. In: Proceedings of ACM Symposium on Applied Computing, pp. 110–119 (2018). https://doi.org/10.1145/3167132.3167141
4. ADONIS Business Process Modelling Suite, BOC Group. https://www.boc-group.com/en/adonis/. Accessed 29 Nov 2021
5. CATIA No Magic website. https://www.3ds.com/products-services/catia/products/no-magic/
6. Santos, P.S., Almeida, J.P.A., Guizzardi, G.: An ontology-based semantic foundation for ARIS EPCs. In: Proceedings of ACM Symposium on Applied Computing, pp. 124–130 (2010). https://doi.org/10.1145/1774088.1774114
7. IDEF – Integrated DEFinition Methods (IDEF). https://www.idef.com/. Accessed 29 Nov 2021
8. Defining and Simulating ITIL Processes Using IBM WebSphere Business Modeler Advanced. https://statemigration.com/defining-and-simulating-itil-processes-using-ibm-websphere-business-modeler-advanced/. Accessed 29 Nov 2021
9. Kalnins, A., et al.: Business Modeling Language GRAPES-BM and Related CASE Tools. http://www.gradetools.com/grade40/white/bmlpiln.htm
10. IDEF0 Overview for Business Process Modeling. https://www.slideshare.net/EdJohnson22/idef0-overview-for-business-process-modeling. Accessed 29 Nov 2021
11. Maull, R., Weaver, A., Smart, A., Childe, S.: Using IDEF0 to develop generic business process models. In: Plonka, F., Olling, G. (eds.) Computer Applications in Production and Engineering. ITIFIP, pp. 227–236. Springer, Boston, MA (1997). https://doi.org/10.1007/978-0-387-35291-6_20

12. WebSphere Business Modeler. http://www.bptrends.com/surveys/07-2007%207-BP%20M
    odeling%20Report%20-%20IBM.pdf
13. Strazdina, R., Kirikova, M.: Business process modelling perspectives analysis. In: Stirna, J.,
    Persson, A. (eds.) PoEM 2008. LNBIP, vol. 15, pp. 210–216. Springer, Heidelberg (2008).
    https://doi.org/10.1007/978-3-540-89218-2_16
14. García-Domínguez, A., Marcos, M., Medina, I.: A comparison of BPMN 2.0 with other
    notations for manufacturing processes. In: AIP Conference Proceedings, vol. 1431, p. 593
    (2012). https://doi.org/10.1063/1.4707613
15. Pereira, J.L., Silva, D.: Business process modeling languages: a comparative framework. In:
    Rocha, Á., Correia, A., Adeli, H., Reis, L., Mendonça Teixeira, M. (eds.) New Advances
    in Information Systems and Technologies. Advances in Intelligent Systems and Computing,
    vol. 444. Springer, Cham (2016). https://doi.org/10.1007/978-3-319-31232-3_58
16. Rosemann, M., Recker, J., Indulska, M., Green, P.: A study of the evolution of the represen-
    tational capabilities of process modeling grammars. In: Dubois, E., Pohl, K. (eds.) CAiSE
    2006. LNCS, vol. 4001, pp. 447–461. Springer, Heidelberg (2006). https://doi.org/10.1007/
    11767138_30
17. Bee-Up for Education. https://www.omilab.org/activities/bee-up.html
18. Liu, C., Zeng, Q., Cheng, L., Duan, H., Cheng, J.: Measuring similarity for data-aware business
    processes. IEEE Trans. Autom. Sci. Eng. 19(2), 1070–1082 (2021)
19. Kirikova, M.: Flexibility of organizational structures for flexible business processes. In:
    Proceedings of the 6th Workshop on Business Process Modelling and Support, pp. 83–90
    (2005)
20. Businska, L., Supulniece, I., Kirikova, M.: On data, information, and knowledge representa-
    tion in business process models. In: Pooley, R., Coady, J., Schneider, C., Linger, H., Barry,
    C., Lang, M. (eds.) Information Systems Development, pp. 613–627. Springer, New York
    (2013). https://doi.org/10.1007/978-1-4614-4951-5_49
21. Lundqvist, M., Sandkuhl, K., Seigerroth, U.: Modelling information demand in an enterprise
    context: method, notation, and lessons learned. Int. J. Inf. Syst. Model. Des. (IJISMD) 7(1),
    75–95 (2011)

# Towards AI-Enabled Assistant Design Through Grassroots Modeling: Insights from a Practical Use Case in the Industrial Sector

Hitesh Dhiman[1(✉)], Michael Fellmann[2], and Carsten Röcker[1,3]

[1] TH OWL University of Applied Sciences and Arts, Lemgo, Germany
{hitesh.dhiman,carsten.roecker}@th-owl.de
[2] University of Rostock, Rostock, Germany
michael.fellmann@uni-rostock.de
[3] Fraunhofer IOSB INA, Lemgo, Germany

**Abstract.** Process modeling is used to understand a business process and to document requirements, but is mostly formalized and limited to modeling experts. This can be a problem when designing interactive systems that incorporate AI elements, since the design can fail to take into account tacit knowledge and context-specific requirements of people that execute the process. While recent discourse has highlighted this gap and called for an exploration into light-weight, grassroots modeling techniques that can be used to model everyday work, it is still unclear how these can be harnessed to design information systems that support work. The aim of this paper is to showcase how a triangulated approach combining three different perspectives - grassroots modeling, theoretical grounding, and first person media, can be used to collaboratively model an informal work activity and design an AI-enabled system to instruct novices to perform that activity. Our experience confirms the assertion that, when provided with the necessary scaffolding, experts without any formal modeling experience can be supported to model their specific, local activities and, in doing so, contribute valuable knowledge to the design of information systems.

**Keywords:** Grassroots modeling · Worker assistant · Worker assistance system · Activity theory · Event storming · Artificial intelligence · Information systems

## 1 Introduction

Enterprise Modelling (EM) and Business Process Modelling (BPM) are well-established methods to model professional and technical processes. However, these approaches are limited in several ways. For one, in focusing on the process from a business stakeholder perspective, they ignore local practices or methods

© The Author(s), under exclusive license to Springer Nature Switzerland AG 2022
E. Nazaruka et al. (Eds.): BIR 2022, LNBIP 462, pp. 96–110, 2022.
https://doi.org/10.1007/978-3-031-16947-2_7

which may also be understood via modeling [28]. Two, the practice of formal modeling is often limited to a few people in the organization who are well-versed with digital modeling tools and methodologies [27].

These drawbacks have crucial consequences when it comes to the eliciting software requirements for technological solutions which incorporate artificial intelligence (AI) elements to support everyday activities in an organization. The design of AI based systems comes with numerous challenges in requirements elicitation and design [33]. It is well known in Software Engineering that the success of any system lies in capturing the domain activity as viewed by experts in the domain, as well as the potential user's context [12]. While the users of such systems may be domain experts, it is highly unlikely that they are also professional modelers, which means that designers face the difficulty of comprehending *where* and *how* experts see the potential for AI support, and, in which way to best implement this support. To solve this problem, techniques are needed that bring together diverse domain experts and system designers in an informal, collaborative setting with the common aim of modeling the business activity and facilitating knowledge exchange.

As a response to this challenge, in the past few years, several informal, grassroots approaches to process and domain modeling have been proposed, for instance the work-system-modeling method [3] in the Information Systems domain, and other collaborative domain-driven methods such as event storming or domain storytelling [13,19] in the Software Development domain [6,12,32]. However, the potential of these techniques to model everyday work activities and in order to develop systems incorporating AI has not yet been explored.

In this paper, we report on how we leveraged a modified form of event storming to support grassroots modeling in the process of designing an assistant for supporting workers in the food processing domain. We enhanced the modeling approach with theoretical grounding and recorded audio/video materials to better understand the expert's domain, train our AI models and finalize the system design. We discuss our experiences and weigh on the benefits and limitations of our approach.

## 2    Related Work

### 2.1    AI and Its Role in Future Workplaces

AI continues to gain widespread acceptance in organizations and the industrial sector, where it is seen integral to the business strategy [30]. Although the definition of AI continues to evolve, at this point in time, most systems in organizations are deploying AI solutions which are focused on a specific, narrow task. This entails the deployment of learned models which are trained on a specific data set chosen to represent the problem domain. Given the fact that AI is still developed and maintained by humans, the generated models reflect human decisions and biases [4].

Owing to several technological and sociodemographic challenges, the applications of AI in the industrial sector are many, both in a *substituting* and *complementing* role [2]. In the latter sense, solutions range from decision-making support [9] to task support in the form of intelligent worker assistants [11]. Findings show that AI solutions can be both beneficial [5,24] or detrimental to work [15]. Which of these outcomes are achieved depends on how the AI solutions are designed, deployed, and maintained.

Hence, when it comes to design, the development of such systems is far from trivial [33]. AI developers and domain experts may come from different domains which makes communication difficult [25]. Moreover, AI models are seldom fully accurate, and may in fact never be complete, depending on the context of the problem. There will be rough edges which require that the interaction with the system is adjusted to reflect these limitations. As [34] states, "it is the developers' understanding, not the domain experts' understanding, that gets released to production". Therefore, if the domain experts and developers share the same mental model (a mental representation of a process, its working, the relationships between its components, and an understanding of the possible range of actions and their consequences on the process), several negative consequences, ranging from missed expectations and acceptance issues can be potentially detected and resolved during the development phase.

Therefore, collaboratively understanding the domain expert's activity and developing a shared mental model is important – firstly, to understand the domain activity (since two domain experts from different domains may have a divergent view of the same activity), secondly, to negotiate expectations and the possible impact of an AI feature on potential users, and thirdly, to design the user-system interaction around the limitations of the AI solution.

## 2.2 Approaches to Modeling and Comprehending Informal Practices

Business process models can be used to capture recurring structures and sequences in a business process via data and control flows. These formal processes or workflows can be used to prescribe the what, how, and who of work. However, the final execution of activities by humans always takes place in a particular work context. In many instances, humans perform activities based on their own skills, experiences and knowledge, processes which do not follow a pre-defined business logic, and can be labeled as 'informal', since they consist of informal practices (e.g. best practices, instructions and training) [23]. It may not be possible to capture these using formal modeling tools for several reasons, among them, the unclear structure of the activity, high time (and therefore, cost) investments, the variability of the activity itself, and high rigor requirements of formal models [31]. While notations such as Case Management Model and Notation (CMMN), have been proposed as an option to capture unstructured activities in the domain of knowledge work, they are designed to be used by modeling experts rather than end users [18].

When developing IT systems for non-IT experts, developers encounter two major issues - first, their own understanding of the problem domain is limited, and second, when designing systems that are to support users in their everyday activities which also include an informal component, relying on formal process models can lack the necessary detail to understand user requirements in context, and in several cases, can lead to translation issues [34]. For this reason among others, in the past few years there has been a call for:

1. Finding collaborative, lightweight methods that enable grassroots modeling [29].
2. Finding/establishing design theories that can form the basis for understanding and modeling both the activity domain and designed artifacts.

Hence, the focus lies on finding ways to collaboratively understand the experts' domain and elicit their knowledge using simplified techniques, thereby creating a shared mental model of the problem domain. We briefly summarize some well-established approaches from three different perspectives - theoretical grounding, first-person viewpoint, and domain modeling.

From a theoretical standpoint, *activity theory* [17] provides a framework to understand the enactment and modeling of informal processes. Based on an established social theory of work, activity theory considers the activity as its basic unit of analysis, consisting of three main entities: the subject (the one performing the activity as an agent), the object (at which the subject directs his/her actions, and which also crystallizes the outcomes of the activity) and the tool (which mediates the actions of the subject). In any work context, the subject and the tool both represent 'internalized' forms of knowledge. Over time, the subject gains experience and learns best practices, while the tool embodies this knowledge in terms of its design and successive improvements [17]. Activity theory has previously been used to model and design information systems [8,14,20].

In addition, a process can also be seen in terms of *events*, defined as "something that happens during the course of a process", a notation well-established in business process modeling. According to *event segmentation theory*, humans intuitively segment events in order to perceive a continuous stream of activity, and this segmentation also scaffolds memory and learning [35]. Events are distinguished hierarchically in terms of fine- or coarse-grained events, with coarse events corresponding to change in objects' features, and fine-grained events relating to actions performed on those objects. These insights, along with the theoretical framework of activity theory, can be used to understand activities in the following manner: Coarse-grained events will refer to the object of the activity, that is, changes in object's states, whereas fine-grained events will apply to the use of the tool including the subject's actions. Viewed this way, while an event signals a change, using activity theory, we can trace this event back to *what* changes, and *how*.

On a related note, the use of recorded media to understand user needs has a long tradition in design methodologies such as user-centered design [7]. However, in the past few years, widespread availability of wearable audio and video recording technologies have made it easier to capture and understand human activities from a first-person point of view [16]. While recorded media has previously been used to automatically recognize human activity or to instruct users [22], its potential role in modeling has not been explored.

Finally, from a practical modeling perspective aimed at system development, *domain driven design (DDD)* is a model-driven approach that attempts to capture the domain activity along with its concepts and relationships by prioritizing the discovery of business events (or *domain events*) which signal a transformation of data rather than focusing on the data structures [34]. DDD requires that domain experts and developers collaboratively create a model of the domain activity based on a shared *ubiquitous language*. One of the approaches to discover events in a DDD context is the workshop format *event storming* [6]. In it, the domain experts and software developers (in a group of 4–8 participants) are invited to write domain events on sticky notes which are posted on a wall. The formal event storming method prescribes a color coded grammar representing different entities in a domain: *event, system, command, policy, read model* and *actors*. Nonetheless, this structure is not enforced rigorously, rather, the aim behind event storming is to reduce barriers and encourage conversations instead of aiming for premeditated precision [6]. The workshop begins with identifying core events in an activity domain, placing them on a timeline, and ends with clarifying the role of entities in triggering, reacting to, or consuming events in the entire process. Once domain-level event storming concludes, the same method can also be applied at a design-level to design a software solution.

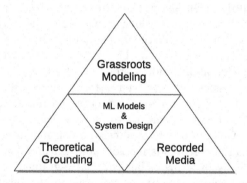

**Fig. 1.** Summary of our overall approach.

The combination of these three approaches (modeling, theory, and recorded media), can be viewed as a *triangulation*, summarized in Fig. 1. Since they view the same process from different perspectives, they can, in theory, complement each other by filling in the gaps created by using each method in isolation. However, in the context of modeling, the use of this triangulated approach has rarely been explored. In the next section, we describe how we combined and used the three perspectives to model an industrial activity with the aim of developing a worker assistant.

# 3   Industrial Use Case

## 3.1   Problem Description

A high degree of digitization and automation notwithstanding, several tasks in the industrial domain are still performed by human workers. Some of these tasks, particularly in domains such as carpentry, food-processing, construction and so on, are considered as skilled, or semi-skilled tasks, since they involve a considerable amount of dexterity and maneuverability backed by training and on-the-job experience. In the European region, many such professions are confronted with a shortage of workers and apprentices due to sociodemographic trends. One way to fill this gap is to technologically assist unskilled workers to learn how take on skilled tasks. Such systems, termed as *worker assistant systems* or *worker assistants*, combine various sensing, analysis and interaction techniques (including AI and mixed/augmented reality), to track activities in order to guide, inform, and teach workers on-the-job. The field of intelligent assistants is not new [11], however, in the past few years, investigating the feasibility of such assistants in various industrial domains has gathered much research interest [21].

Our particular problem here concerns the food-processing sector, that is, meat processing. The activity here involves separating a pork shoulder into various cuts of meat, and is usually performed by an expert with at least 3–6 years of training in the domain. Our task here was to design an AI-enabled assistant that can guide a novice worker to perform the same activity. From the perspective of worker assistant design, the domain is of interest because no existing examples of assistants in this domain can be found. In the next sections, we describe how we gathered domain knowledge, modeled the domain activity and designed the assistant.

## 3.2   Approach

Figure 2 illustrates our design process. The overall process consisted of two phases. We carried out the modeling activities in the first phase to gather domain knowledge. In the second phase, we focused on modeling the system.

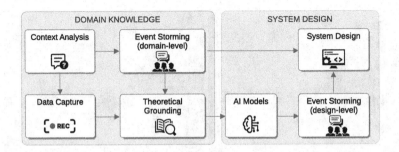

**Fig. 2.** Design process followed in the project.

**Context Analysis.** In the first step (termed *context analysis*), we consulted a domain expert (with 6 years of training in the domain) to understand how the activity is carried out generally in terms of regulations and constraints (both physical and environmental) influencing the activity. From a formal process perspective, the activity is standardized, and described as follows:

1. Either leave the pork knuckle as a whole or prepare it for further processing by removing the forearm bone
2. Cut the joint between the scapula (shoulder-blade) and humerus (leg) bones, pre-cut the bones on both sides
3. Pull the shoulder blade with periosteum from the scapula bone
4. Expose the joint on the scapula, pull the bones cleanly, completely remove the cartilage

It is apparent that the description above is relatively brief. Cutting, deboning, and filleting is a manual process performed using a stainless steel knife. The entire sequence entails a prior understanding of pig anatomy, the sequence of steps, and various methods of cutting. The steps can be performed out-of-step if needed. Unlike manufactured parts that are produced to be exactly identical, animal parts have natural variations. The cutting process must take these into account.

**Event Storming (Domain-Level).** In order to deepen our understanding of the activity, we invited the expert to an event storming workshop. The expert had no prior knowledge of process modeling or digital modeling tools. Deviating from the typical, 'business oriented' focus of event-storming workshops, we instead requested the expert to write down the 'events', or 'things that happen' during this activity on sticky notes and place them in a temporal sequence. The activity sequence as explained by the expert is shown in Fig. 3. In addition to posting the events, the expert also insisted on adding additional information to those events (marked in Fig. 3 with 'i'). These can be categorized into three classes - visual descriptions of the state of the pork shoulder to determine the 'correctness'

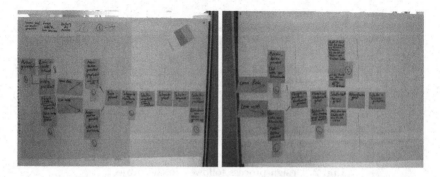

**Fig. 3.** Understanding the activity in an event storming session with the domain expert.

of each step, decisions to be taken (based on the state of the pork shoulder), and tips relating to the quality of work being performed. Doing event storming collaboratively also opened up opportunities for discussions with the expert, e.g. which events could be possibly captured by an AI detection model. The process model obtained here listed the sequence of domain events, but did not contain any information about the actions that lead from one event to the next.

**Data Capture.** To understand the activity dynamics, we conducted a demonstration study with the expert to capture data (Fig. 4). The purpose of doing so was threefold: first, to deepen our own comprehension of the activity. Second, to gather data for segmentation, classification and training of the AI models, and third, to discover features or actions the expert may have neglected to mention due to their tacit nature. We used three different sensors: a forehead mounted GoPro camera (used to capture the overall process flow from a first-person perspective), a gaze tracker (to detect the points on the pork shoulder that the expert focused on while carrying out the activity), and a smartwatch with an in-built accelerometer (to record hand movements). The expert performed the activity ten times, generating about 30 min of video footage and associated accelerometer data.

**Fig. 4.** Snippets from the data capture study. The three photos (from left to right) also show the three different ways in which the expert uses the knife.

**Theoretical Grounding and Step Ontology.** After gathering the data, we combined the insights from the event storming session and theoretical grounding to create a step model (Fig. 5) which combines the principles of activity theory and event segmentation theory with the description of steps as explained by the expert. Hence, the boundaries between main events posted by the expert in the event storming session ought to correspond to coarse changes in features of the object (pork shoulder), and the finer events ought to correspond to actions (hand movements and knife grips) to be found in the videos and sensor data. A particular sequence of hand movements and grip can be subordinated to each step, or sub-step, which are representative of the state of a particular step.

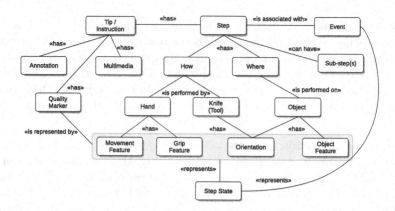

**Fig. 5.** Step-model based on theoretical grounding and domain knowledge.

**Creation of AI Models.** To train AI models which could follow user activity, we classified the visual data into coarse events, and used them to train an object classifier (YoloV3 [26]). We chose features corresponding to the events mentioned by the expert. In step 1, meat covering the inner part of the shoulder is first removed, followed by separating the leg. In step 2, the skin on the reverse side is removed. During steps 3 and 4, the expert cuts through the meat covering and surrounding the humerus (foreleg) bone, exposing it and the joint connecting it to the scapula (shoulder) bone. The joint is cut and the scapula is then separated from the underlying meat, followed by the humerus bone. From an event segmentation perspective, while we were able to recognize several events corresponding to parts of the shoulder that are removed, the AI classifier could identify four unique features - the leg that is separated in step 1, the skin removed in step 2, and the two bones that are removed in steps 3 and 4. All other parts of the pork shoulder are not classified correctly since they are fairly homogeneous in their texture.

Next, regarding fine events corresponding to sub-steps, during the event storming session the expert drew attention to the different ways in which the knife is to be held, but found them hard to communicate in a written format. Hence, we looked for uniquely determinable hand movements or ways of holding the knife in both the visual and sensor data. We found three distinct knife movements (cutting, carving, and slicing sideways) and three different knife grips (forward grip, reverse grip, and forward thumb support grip). To detect these hand grips and movements we trained a model (using TensorFlow [1]) which runs separately on an android smartphone connected to the smartwatch.

The complete process of data classification and training is explained elsewhere [10]. Our initial aim was to train the AI models to autonomously apply the event boundary model as envisioned in Fig. 6 to support continuous activity tracking. However, the limitations encountered here meant that the user would have to indicate to the system the step being executed.

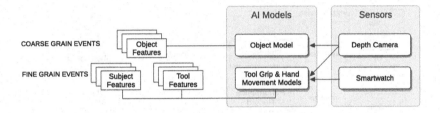

**Fig. 6.** Creation of AI models.

At this point, we had: (1) a model of the activity according to the expert (2) additional data in the form of multimedia that can be collated with the expert's explanations, instructions and tips, and (3) AI models which could be used to track some features of the user's activity.

**Event Storming (Design-Level).** As a final modeling step, we organized a second event storming workshop to design the new process of working together with the assistant, this time involving only the developers. Again, the intention behind the workshop was to foster problem-solving discussions and discover additional domain entities or aggregates (a group of entities) including the ones already known, using the shared mental model and video data gathered earlier.

The workshop was held in three sessions in the same format as earlier, using the event storming notation. The developers started with posting events that would be triggered directly by the user, and indirectly by AI models. This is a deviation from the usual event storming flow, since in it, an entity (an aggregate or an external system) usually responds to a command triggered by the user or a policy. In our case, we treated the AI models as 'autonomous' since they are tracking the user's activity. Both the event storming models were transferred to a digital format using an online diagramming tool (LucidChart).

**System Design.** After the session ended, we demarcated the domain aggregates and their associated events. Two aggregates were derived directly from the user's activity - the pork shoulder and the user's hand. In addition, a 'step store' aggregate was added as the repository responsible for supplying the basic instructions/hints provided by the expert, multimedia resources and reference object/hand/tool features. A 'performance record' entity was added to store the user's performance data during each step (e.g. the deviation of the user's hand movements and knife grip from the expert's). The two AI models developed earlier were included as systems (hand gesture model and the shoulder/object detection model). Policies can be triggered as either corrective measures to inform the user, or as general system level decisions to start or stop the detection models.

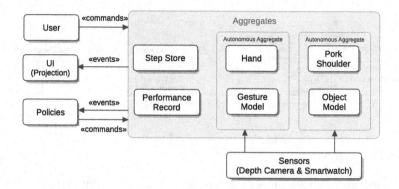

**Fig. 7.** Event-driven architecture of the prototypical assistant.

The final, event driven system design for the assistant is shown in Fig. 7. The four final aggregates and the two AI models can receive commands from the user, and in changing the nature of the aggregates, issue events which can either be directly routed to the User Interface (UI), or invoke specific policies which then, in turn, issue commands to the aggregates. This model serves as the final design artifact before beginning with the process of prototyping. Based on our context analysis, for the presented use case, we decided to use a projection based system in which the UI is projected onto the worktable. A depth camera captures both color and depth information. The color data is fed to the object detection model as well as the hand recognition model, whereas the depth data is used to enable interactivity with the projected UI.

## 4  Discussion

In our article, we have demonstrated how 'grassroots model creation' can be used to elicit expert knowledge and guide the design of an artifact (an assistant) that supports existing work. By taking a more hands-on approach and excluding the use of prior notations and digital tools, the initiation of the modeling process was greatly accelerated. As an example, in our case the expert had no knowledge of event storming as such, but was able to quickly put events and associated information on paper. While this confirms the existing view that 'people do modeling without even noticing it' [29], it also shows that people may require some scaffolding to structure the process. Techniques such as event storming [6] or domain story telling [13] can provide valuable starting points and be adapted to a particular context with relaxations or deviations.

In addition, using additional data in the form of photos and videos proved to be an extremely helpful resource in understanding the expert's work and modeling it. In fact, the videos revealed additional actions that the expert performed but did not mention during our event storming session, most likely because these were tacit in nature. Although these techniques have been mostly limited to user experience design (e.g. in the form of field observation), with reduction in size and costs of video/audio recording technology, the barrier to using these techniques has been greatly reduced. While not mentioned in previous discussions on grassroots modeling, our work adds this new dimension of using first-person records to enrich modeling, and demonstrates the importance of these additional resources.

Further, by basing our modeling activity in activity theory and event segmentation theory, we have also shown how these two could enhance grassroots modeling. More specifically, our example touches on the sociological aspect of 'studying practices'. The use of activity theory is already well-established in the fields of Human-Computer Interaction (HCI) or Computer Support Cooperative Work (CSCW), the use of event-segmentation theory not so much. In our case, the theoretical perspective lends a formal basis for defining and identifying activity entities and events, leading to a shared vocabulary as envisioned in DDD.

As Fig. 1 summarizes, our approach is a form of a *triangulation*. No one perspective provides us with the complete picture, hence it is helpful to combine different approaches. Event storming as a method of scaffolded grassroots modeling explicates expert knowledge, but may hide tacit knowledge. Here the record of expert activity captured from a first-person perspective fills in the gaps. In addition, there is no fixed starting point to this approach - one could begin with either of the three perspectives and gradually enrich one's knowledge by bringing in other aspects.

Finally, our contribution shows how the role of AI in work can be better clarified in a collaborative manner with an existing model of the workflow and associated work entities. More importantly, our approach shifts the goal of using AI from a purely business perspective to a more worker-centric vantage point. Collaboratively modeling AI solutions exposes their bottlenecks to both the developers and domain experts and benefits both. For domain experts, it allows them to choose which aspects of their work can be supported autonomously and which be kept manual. For developers, it generates insights on how to best integrate these solutions into an interactive system.

## 5   Limitations and Future Work

This work also has its limitations. For one, the use of a non-digital format means that there is no easy way to directly translate the event model into an existing digital workflow. We think this is due to the underlying motivation behind event storming, to quickly move from an understanding of the business domain to software design. We converted the results of the event storming model into a digital format using an online tool (LucidChart). In any case, the work of creating a step ontology and software design is still left to the developers.

One could also argue that the scope of modeling in this work is rather narrow, and focuses on only a particular activity. However, as Figs. 5 and 6 show, the resulting step model and approach is generic in nature. By grounding our work in existing theory, we are of the opinion that the step model can be extended to several activities and domains, both as a support for modeling and as an initial artifact serving as a starting template for the development process. As for our approach to creating AI models, event segmentation theory has already spurred efforts in automated event boundary detection to separate a continuous activity into chunks, and hence steps. It would be interesting to explore the use of such automated techniques in modeling.

From the perspective of AI, we only focused on the limited data set gathered for this particular use case. Creating a robust AI model requires training and classifying a large amount of data, and one way in which this could be achieved is by empowering domain experts to refine and fine-tuning the models later. Doing so would require additional collaboration with domain experts and design of human-centric approaches.

## 6    Conclusion

The vision of grassroots modeling has emphasized the role of people in the modeling process. However, in several industrial domains, work activities are rarely modeled, although they rely on well-established norms and conventions gathered via experience. Eliciting these perceptions and motivations is necessary when designing systems intended to support people at work. In our paper, we have investigated one such approach to grassroots modeling by using event storming as a light-weight, informal modeling technique to design a prototypical AI-based worker assistant in the food-processing sector. In our use-case, event storming provided the starting point to understand work, and we enriched this activity model with additional media (e.g. videos). We relied on activity theory and event segmentation theory to extract features for both classifying and training our AI models and to design an event driven architecture. We hope that our work provides opportunities for further research and discussion.

## References

1. Abadi, M., et al.: TensorFlow: large-scale machine learning on heterogeneous systems (2015). https://www.tensorflow.org/, software available from tensorflow.org
2. Acemoglu, D., Restrepo, P.: The race between man and machine: implications of technology for growth, factor shares, and employment. Am. Econ. Rev. **108**(6), 1488–1542 (2018). https://doi.org/10.1257/aer.20160696
3. Alter, S.: The work system method for understanding information systems and information systems research. Commun. Assoc. Inf. Syst. **9**(1), 6 (2002)
4. Barredo Arrieta, A., et al.: Explainable artificial intelligence (XAI): concepts, taxonomies, opportunities and challenges toward responsible AI. Inf. Fusion **58**, 82–115 (2020). https://doi.org/10.1016/j.inffus.2019.12.012, https://linkinghub. elsevier.com/retrieve/pii/S1566253519308103
5. Brachten, F., Brünker, F., Frick, N.R.J., Ross, B., Stieglitz, S.: On the ability of virtual agents to decrease cognitive load: an experimental study. Inf. Syst. e-Bus. Manage. **18**(2), 187–207 (2020). https://doi.org/10.1007/s10257-020-00471-7

6. Brandolini, A.: Introducing Event Storming: An Act of Deliberate Collective Learning. LeanPub (2017)
7. Brun-Cottan, F., Wall, P.: Using video to re-present the user. Commun. ACM. **38**(5), 61–71 (1995). https://doi.org/10.1145/203356.203368
8. Collis, B., Margaryan, A.: Applying activity theory to computer-supported collaborative learning and work-based activities in corporate settings. Educ. Technol. Res. Dev. **52**(4), 38–52 (2004). https://doi.org/10.1007/BF02504717
9. Dellermann, D., Lipusch, N., Ebel, P., Leimeister, J.M.: Design principles for a hybrid intelligence decision support system for business model validation. Electron. Mark. **29**(3), 423–441 (2019). https://doi.org/10.1007/s12525-018-0309-2
10. Dhiman, H., Alam, D., Qiao, Y., Upmann, M., Röcker, C.: Learn from the best: harnessing expert skill and knowledge to teach unskilled workers. In: The15th International Conference on PErvasive Technologies Related to Assistive Environments, pp. 93–102. ACM, Corfu, Greece, June 2022. https://doi.org/10.1145/3529190.3529203
11. Dhiman, H., Wächter, C., Fellmann, M., Röcker, C.: Intelligent assistants: conceptual dimensions, contextual model, and design trends. Bus. Inf. Syst. Eng. (2022). https://doi.org/10.1007/s12599-022-00743-1
12. Evans, E.J.: Domain-Driven Design: Tackling Complexity in the Heart of Software. Addison-Wesley Professional, Boston (2004)
13. Hofer, S., Schwentner, H.: Domain Storytelling: A Collaborative, Visual, and Agile Way to Build Domain-Driven Software. 1st Edn. The Addison-Wesley Signature Series/a Vaughn Vernon Signature Book, Addison-Wesley, Boston (2021)
14. Iyamu, T.: A case for applying activity theory in IS research. Inf. Resour. Manage. J. **33**(1), 1–15 (2020). https://doi.org/10.4018/IRMJ.2020010101
15. Johnson, A., et al.: A review and agenda for examining how technology-driven changes at work will impact workplace mental health and employee well-being. Austr. J. Manage. **45**(3), 402–424 (2020). https://doi.org/10.1177/0312896220922292
16. Kanade, T., Hebert, M.: First-person vision. Proc. IEEE. **100**(8), 2442–2453 (2012). https://doi.org/10.1109/JPROC.2012.2200554, http://ieeexplore.ieee.org/document/6232429/
17. Kaptelinin, V., Nardi, B.A.: Acting with Technology: Activity Theory and Interaction Design. MIT Press; Ebsco Publishing, Cambridge; Ipswich (2007). http://ieeexplore.ieee.org/servlet/opac?bknumber=6267290
18. Kurz, M., Schmidt, W., Fleischmann, A., Lederer, M.: Leveraging CMMN for ACM: examining the applicability of a new OMG standard for adaptive case management. In: Proceedings of the 7th International Conference on Subject-Oriented Business Process Management, pp. 1–9. ACM, Kiel Germany, April 2015. https://doi.org/10.1145/2723839.2723843
19. Lai, H., Peng, R., Ni, Y.: A collaborative method for business process oriented requirements acquisition and refining. In: Proceedings of the 2014 International Conference on Software and System Process - ICSSP 2014, pp. 84–93. ACM Press, Nanjing, China (2014). https://doi.org/10.1145/2600821.2600831
20. Luukkonen, I., Korpela, M., Mykkänen, J.: Modeling approaches in the early phases of information systems development. In: Proceedings of the 18th European Conference on Information Systems (2010)
21. Mark, B.G., Rauch, E., Matt, D.T.: Worker assistance systems in manufacturing: a review of the state of the art and future directions. J. Manuf. Syst. **59**, 228–250 (2021). https://doi.org/10.1016/j.jmsy.2021.02.017, https://linkinghub.elsevier.com/retrieve/pii/S0278612521000534

22. Molino, A.G., Tan, C., Lim, J.H., Tan, A.H.: Summarization of egocentric videos: a comprehensive survey. IEEE Trans. Hum. Mach. Syst. 1–12 (2016). https://doi.org/10.1109/THMS.2016.2623480, http://ieeexplore.ieee.org/document/7750564/

23. Moody, P., Gruen, D., Muller, M.J., Tang, J., Moran, T.P.: Business activity patterns: a new model for collaborative business applications. IBM Syst. J. **45**(4), 683–694 (2006). https://doi.org/10.1147/sj.454.0683, http://ieeexplore.ieee.org/document/5386642/

24. Nazareno, L., Schiff, D.S.: The impact of automation and artificial intelligence on worker well-being. Technol. Soc. **67**, 101679 (2021). https://doi.org/10.1016/j.techsoc.2021.101679, https://linkinghub.elsevier.com/retrieve/pii/S0160791X21001548

25. Piorkowski, D., Park, S., Wang, A.Y., Wang, D., Muller, M., Portnoy, F.: How AI developers overcome communication challenges in a multidisciplinary team: a case study. Proc. ACM Hum. Comput. Interact. **5**(CSCW1), 1–25 (2021). https://doi.org/10.1145/3449205

26. Redmon, J., Farhadi, A.: YOLOv3: An Incremental Improvement. arXiv:1804.02767 [cs], April 2018

27. Ross, J.W., Quaadgras, A.: Enterprise architecture is not just for architects. CISR Res. Brief. **7**(9), 1–9 (2012)

28. Sandkuhl, K., et al.: Enterprise modelling for the masses – from elitist discipline to common practice. In: Horkoff, J., Jeusfeld, M.A., Persson, A. (eds.) PoEM 2016. LNBIP, vol. 267, pp. 225–240. Springer, Cham (2016). https://doi.org/10.1007/978-3-319-48393-1_16

29. Sandkuhl, K., et al.: From expert discipline to common practice: a vision and research agenda for extending the reach of enterprise modeling. Bus. Inf. Syst. Eng. **60**(1), 69–80 (2018). https://doi.org/10.1007/s12599-017-0516-y

30. Shrestha, Y.R., Ben-Menahem, S.M., von Krogh, G.: Organizational decision-making structures in the age of artificial intelligence. California Manag. Rev. **61**(4), 66–83 (2019). https://doi.org/10.1177/0008125619862257

31. Sungur, C.T., Binz, T., Breitenbucher, U., Leymann, F.: Informal process essentials. In: 2014 IEEE 18th International Enterprise Distributed Object Computing Conference, pp. 200–209. IEEE, Ulm, Germany, September 2014. https://doi.org/10.1109/EDOC.2014.35, http://ieeexplore.ieee.org/document/6972068/

32. Vernon, V.: Implementing Domain-Driven Design. Addison-Wesley, Boston (2013)

33. Wan, Z., Xia, X., Lo, D., Murphy, G.C.: How does machine learning change software development practices? IEEE Trans. Softw. Eng. 1 (2020). https://doi.org/10.1109/TSE.2019.2937083, https://ieeexplore.ieee.org/document/8812912/

34. Wlaschin, S., MacDonald, B.: Domain modeling made functional: tackle software complexity with domain-driven design and F#. The pragmatic programmers, The Pragmatic Bookshelf, Raleigh, North Carolina, version: p1.0 edn. (2018)

35. Zacks, J.M., Swallow, K.M.: Event segmentation. Curr. Direct. Psychol. Sci. **16**(2), 80–84 (2007). https://doi.org/10.1111/j.1467-8721.2007.00480.x

# Applications and Technologies

# Analytics in Industry 4.0: Investigating the Challenges of Unstructured Data

Michael Möhring[1][(✉)], Barbara Keller[3], Rainer Schmidt[2], Fabian Schönitz[3], Frederik Mohr[3], and Max Scheuerle[3]

[1] Faculty of Computer Science - HHZ,, Reutlingen University, Alteburgstraße 150, 72762 Reutlingen, Germany
`research@michael-moehring.de`

[2] School of Computer Science and Mathematics, Munich University of Applied Science, Lothstr. 35, 80335 Munich, Germany

[3] DHBW Stuttgart, Business Information Systems, Cooperative State University Baden-Wuerttemberg, Paulinenstr. 50, 70178 Stuttgart, Germany

**Abstract.** Data analysis is becoming increasingly important to pursue organizational goals, especially in the context of Industry 4.0, where a wide variety of data is available. Here numerous challenges arise, especially when using unstructured data. However, this subject has not been focused by research so far. This research paper addresses this gap, which is interesting for science and practice as well. In a study three major challenges of using unstructured data has been identified: analytical know-how, data issues, variety. Additionally, measures how to improve the analysis of unstructured data in the industry 4.0 context are described. Therefore, the paper provides empirical insights about challenges and potential measures when analyzing unstructured data. The findings are presented in a framework, too. Hence, next steps of the research project and future research points become apparent.

**Keywords:** Unstructured data · Industry 4.0 · Data analytics · Information systems · Data science

## 1 Introduction

The importance of data as a strategic asset becomes increasingly dominant from day to day [1]. However, not all data is equal. On closer inspection, it can be seen that up to 90% of the collected data today is unstructured data like textual, audio or video data [2]. Experts assume that this data also contains an abundance of valuable information that can be made available with the help of analytics and used for data-driven optimization of processes, especially in industry [3]. This sounds simple at first, but companies must first overcome a number of hurdles to access this value. One point is, that the analysis of unstructured data is quite difficult for companies because of a missing defined structure coming up with no related underlying conventional data models [2]. It is therefore not surprising that only a minority of companies are currently able to analyze unstructured

Ē. Nazaruka et al. (Eds.): BIR 2022, LNBIP 462, pp. 113–125, 2022.
https://doi.org/10.1007/978-3-031-16947-2_8

data and reap the associated benefits [2, 3]. However, the utility and value of analyzing unstructured data remains undisputed and can be illustrated by many application areas. For instance, in terms of predictive maintenance [32]. Images can be used to detect defects or critical wear at individual points on the production line. In this way downtimes can be prevented, or setup times can be scheduled. Here, the value of analyzing unstructured data becomes very clear, as not all wear can be quantified.

A context, which is particularly affected by the potentials and challenges of data analytics and business intelligence is industry 4.0 [4]. The term industry 4.0 embraces several concepts. Here, the definition of Lasi et al. [5] defining industry 4.0 embracing the concepts smart factory, cyber-physical systems and self-organization is used. Furthermore, industry 4.0 introduces new systems in distribution and procurement and in the development of new products and services [5]. Analytics and business intelligence are of increasing importance for industry 4.0 [4]. On the technical level the integration of industry 4.0 and Big Data is increasingly discussed [6, 7]. Manufacturing enterprises already started to use unstructured data for analyses in the area of Industry 4.0 and information systems research is investigating these challenges (e.g., in [8, 9]). However, a literature review (Sect. 2) following the recommendations of Webster and Watson [10] shows that there is sparse research on the challenges of the usage of unstructured data for industry 4.0 applications. For instance, only a few authors identified the usage of unstructured data in Industry 4.0 (e.g., [13, 33, 34]) but without deep empirical evidence (Sect. 2). Therefore, this paper as part of a collaborative research project addresses the research question: *"What are the challenges of the usage of unstructured data for industry 4.0 related analytics?"*.

The paper starts in the following section with insights into the research background, which are derived from a comprehensive systematic literature review. Then our study is presented by describing the method, the data collection and data analysis. This is followed by a presentation of the research findings. The developed initial framework for the collaborative research project is shown and the results of the study contributing to answer the research question on a first step are described. Finally, the paper ends with a conclusion and discussion section, where also future research steps are provided.

## 2 Related Research and Research Background

Aiming to investigate the research question's subject a systematic literature review according to the guidelines proposed by Webster & Watson [10] was conducted. In a first step, a literature search was performed. Therefore, initially terms such as "unstructured data", "industry 4.0", "analytics" were insert into major research database platforms (e.g., AISeL, SpringerLink, ScienceDirect, IEEE Xplore) and a search was conducted. Afterwards, they were additionally complemented by synonyms such as "smart manufacturing" or "big data". This procedure revealed about 30 sources. Within the framework of selection, the focus was set on reputational journals as well as leading conferences. For this reason, the quality of the research and the importance of the findings can be assumed. Each paper was reviewed carefully, and the attention was especially paid on the abstract, the methodology, the provided results as well as the derived implications. Hence, the findings of different studies assumed to contribute to the raised research question are presented comprehensively in the following. The literature review revealed that only few researchers identified the importance of data and in particular unstructured data in their studies.

Tao and colleagues [33] discussed in their study the concept of data-driven smart manufacturing. They created a phase model in which structured and unstructured data is used for different purposes. Starting point is the smart design phase that uses data to support market analysis and demand analysis. Especially, unstructured data created by customers' reviews etc. plays an important role. In the next phase, production planning and process optimization, primarily structured data is used for. The following phase, material distribution and tracking, uses unstructured data for, tracking material, supervising material supply and detecting quality deficiencies. In sum, authors provided the insight that unstructured data also helps during manufacturing and process monitoring to track the progress of manufacturing and to detect interruptions. Production quality control intensively uses unstructured data e.g., images to determine the quality of production output. Smart equipment maintenance uses images, videos and audio files created from devices in operation to predict faults and optimize energy consumption.

The use of unstructured data for deep learning in smart manufacturing is in the focus of the research of Wang and colleagues [34]. First, they identify sources of unstructured data in smart manufacturing such as objects, equipment, product, people and the environment. Then they describe the use of unstructured data for descriptive, diagnostic, predictive and prescriptive purposes and derive different recommendations. They propose the use of unstructured data for capturing products' condition, environment and operation to understand what happened (descriptive analytics). For diagnostic purposes, that means to understand why something happened, the authors suggest using unstructured data to examine the causes of reduced product performance or detect failure. Beside the investigation of using unstructured data in the backward-looking perspective, they also examined the future-oriented perspective. Here, authors provided insights how unstructured data can help to identify and detect up-coming incidents (predictive analytics) by determining product quality and patterns that signal impending events. In addition, they highlight the use of unstructured data to determine what actions to use when improving outcomes or identifying approaches to address problems (prescriptive analytics).

Investigating the importance of unstructured data, Arnold and colleagues [13] identified and proclaimed its relevance as architectural feature of IIoT (Industrial Internet of things) platforms. As part of a taxonomy of IIoT platforms' architectural features, they associated unstructured data with possible use cases in descriptive, real-time, predictive and prescriptive analytics. The authors also describe the data flow within the architecture. Unstructured data flows into the applications layer where APIs are provided. The unstructured data are collected via the network layer from the infrastructure layer.

Numerous other authors investigated selected aspects data analytics within Big Data, industry 4.0 and related concepts in this fields. Here, the following studies provide insights highlighting the importance to aim especially strategic and managerial goals.

Fay and Kazantsev [17] conducted a study investigating the business value creation mechanisms from Big Data Analytics in smart manufacturing. They start from the observation that in smart manufacturing settings a multitude of heterogeneous data sources providing both structured and unstructured data exist. However, the impact of these new data sources particularly those providing unstructured data, on value creation is not clearly identified so far. Therefore, the authors used a multiple case study research design with semi-structured interviews. Building on those insights, they classified the effects of Big Data Analytics into first- and second order effects and investigate the impact of Big Data Analytics on strategic goals. Results showed that first order effects are on equipment and asset monitoring, especially predictive maintenance scenarios, product quality, and the optimization of the production process. Second order effects are creating by the diffusion of first-order effect to business users, customers and partners. Moreover, the authors found that Big Data Analytics enables new business models, products and services. Big data for cyber physical systems is another aspect discussed in research.

The state of the art in business intelligence in Industry 4.0 is presented by Bordeleau and colleagues [4]. The authors use a systematic review to identify sources of operational and strategic values. Furthermore, the performance indicators found were classified into process-customer, and finance-oriented ones. The authors summarized their findings in a six-layer architectural model for associating the value creation effects of business intelligence with layers of abstraction. In the same way as the research of Bordeleau and colleagues [4], Köhler [23] provides use cases for applying industry 4.0 in practice. Pauli and Lin [27] created a single case study to investigate how actors leverage the generative potential of IIOT platforms. Aiming to operationalize the concept of generativity, the authors assessed the solutions found in two dimensions: first the diversity of solution scenarios and second the capabilities of smart products. In sum, the authors conclude that the generativity of IIoT platforms currently might not be leveraged to its full potential. Likewise, the capability to process even unstructured data in Industry 4.0 environments has been identified by different authors [12, 30]. A high-level research agenda for the internet of things is introduced in [14].

Xu and Duan [36] used a survey and provided comprehensive insights about the state of the art of big data for cyber physical systems. The challenges between big data analytics and cyber physical systems are standing in the focus of Abdulrahman and colleagues [11]. A comprehensive bibliometric analysis of the use of big data analytics and enterprise is provided by Khanra and colleagues [22]. Moreover, the use of industrial big data analytics with cyber-physical systems to enable maintenance and service innovation is discussed in the research of [38]. Li et al. [7] used an inductive qualitative approach to empirically identify the enablers for embedding big data solution in smart factories. Therefore, in semi-structured interviews with experienced consultants and IT managers where conducted. In result, the authors identified in particular integrated management systems as an important enabler for big data in smart factories. The basic concepts of industry 4.0 are discussed in the research of Schmidt and colleagues [37]. Furthermore, research highlighting a conceptual architecture for big data in industry 4.0 can be found [18].

None of the studies mentioned above, however, addresses the gap between unstructured data and data analysis particularly industry 4.0, and the challenges occurring when tying them together. This gap underpins the importance of the central research topic, which is here addressed with a first empirical study as a part of a collaborative research project.

## 3   Method, Data Collection and Data Analysis

*Method:* Aiming to gain insights contributing to the raised research question and the highlighted gap in research, as exploratory research design was chosen. This kind of research design is commonly recommended in cases where an initial understanding and/or a description of a phenomenon must be generated [39]. As a method, a qualitative research approach following the guidelines proposed by Myers and Avison [26] has been chosen. In their guidelines the usefulness of qualitative research in investigating under-researched phenomena is highlighted.

*Data Collection:* As data collection technique, descriptive interviews were conducted. This approach is recommended in literature whenever the phenomenon is to be described from different perspectives, so that the conception is as holistic as possible [39]. The interviews were conducted informal via sending a questionnaire with semi-structured questions to previously be identified experts. The questionnaire contained 14 questions in total. Three questions were assessing the expertise of the participants and another three questions were assessing information about their current employer. These details were integrated into the study as structure variables. The research question itself assessed by 8 open-ended questions focusing different viewpoints of the research question among one explicitly request to list-up challenges of using unstructured data for analysis in the industry 4.0 context. Therefore, side-aspects and the essential point were asked to map the insights and information provided by the experts as well as possible and gain meaningful findings contributing to the research question. Additionally, participants had the possibility to contact the researchers via email. Therefore, we ensured that following-up question or further comments could be also added and discussed like it is common in face-to face settings, in example [39]. All questions and aspects addressed within the questionnaire were discussed and reviewed with in advance with experts not participating in the study. Hence, the understanding of the survey's questions and their meaningfulness was ensured. Furthermore, check questions regarding the experience in the area of data analytics and industry 4.0 were assessed to ensure the expertise of the respondents in the investigated subjects and hence, the quality of the provided information.

*Sample:* The study was conducted in the second quarter of the year 2020. As respondents, experts have been identified by an industry specific search using career networking platforms (e.g., LinkedIN) and personal contacts. In sum, answers from $N = 21$ experts from leading organizations in the area of Industry 4.0 were collected. The companies come from four different sectors. All of the experts were coming from Europe and are also active in this region in a national and international context. The participants coming from organizations with approx. 56000 employees on average in leading manufacturing, IT and consulting enterprises. The experts have a working experience of approx. 19 years on average. Within their organizations a normal data analytics project takes about 6.23 years.

*Data Analysis:* For the analysis of the data a procedure was chosen that follows the rules of coding qualitative data [24]. Following these guidelines, a stepwise data analysis was conducted. In a first step, the content of the qualitative data was reviewed, and codes haven been built. This means, the content was abstracted to different topics. This procedure identified about 40 different codes. In the second step, the codes were reviewed again, and categories have been built by combining different codes addressing similar topics. Finally, the categories were set in relationship to each other. Throughout the entire analysis process quality assurance loops were implemented. In example, coders had to reflect every step of their data analysis in retrospect. This helps to avoid blind spots, in example, and hence it contributes to increasing the quality of the results [15]. The findings of the analysis are presented comprehensively in the framework (Fig. 1) and described in the following [24].

## 4 Results

The data analysis results are shown in the following framework, which is shown in Fig. 1.

As a preliminary condition, all of the experts are agreeing that management from business departments (e.g., product division) as well as related technology and IT departments have to make a decision to run an analytics projects with the use of unstructured data. Although, semi- and unstructured data can be used as data sources in manifold ways, specific projects goals seem to be predominant. In general, experts expect high benefit for *production cost reduction, production quality improvement* as well as *production time optimization.* For instance, running a predictive maintenance project [32] unstructured data sources (e.g., images from the production machine) could be useful to detected reasons for incidents that cannot be quantifiable [29]. One's might think about irregular wear of cutting equipment, which can be detected by picturing the work piece. Furthermore, it could be an *enabler for new service or product innovation.* Additionally, the respondents mentioned the importance of all available data sources.

Our experts are unanimous in their view that a combined analysis of unstructured and structured data (e.g., cycle times, amount of loss) together provides the most value for the organizations:

*"Only when you are able to analyze both in combination the full significance of analyzing unstructured data in an industrial environment does unfold."*

However, the experience of our experts shows that a variety of critical points must be overcome to generate these benefits. Unstructured data in particular pose a major challenge here. The main basic points are: *Analytical Know How, Data Issues and Variety.*

### 4.1 Analytical Know How

Regarding our experts' opinions, many analytical projects are struggling due to missing knowledge and abilities to analyze unstructured data in generals and particularly in the industry 4.0 context. Besides the vast amount of data, that must be handled in this context, the analyses of unstructured data (e.g., images, text, voice) is furthermore quite different to the analyses of structured data (e.g., analyses the amount of produced goods or scrap) [25]. A specific knowledge and experience, how to use and combined different

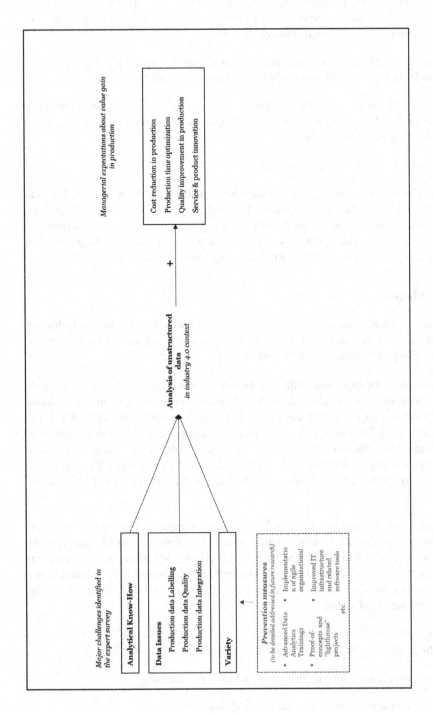

**Fig. 1.** Framework of data analytics challenges of unstructured data for industry 4.0

analytical approaches (e.g., text mining, image recognition), new database concept (e.g., document-based databases) and processing pipelines are needed [20]. These aspects are highlighted by an expert in the following statement:

> *"Large data sets of unstructured data like images/audio require a special know-how in Data Science, which not all Data Scientists are able to provide."*

Other experts are aware of this problem. In this context, they see the development of knowledge and skills within the related departments and disciplines a particular solution, as the following statement proves:

> *"Technology and technology knowledge must be built up"*

In conclusion, the experts provided the insight that the use of unstructured data is often inhibited by the lack of competence to handle and analyze this data. Therefore, organizations should invest in the training of employees or onboard new experts to compensate this lack. This is not only imperative to prevent a failure of the project, but also derive benefits out of analytical projects using unstructured data.

## 4.2  Data Issues

Data issues are another fundamental factor the experts mentioned. Under this factor can be distinguished several aspects. The data analysis provided evidence for the importance of data labelling, data quality and data integration.

*Data Labelling:*  Our experts see different data challenges when it comes to the analyses of unstructured data. First of all, data has to be labelled. For instance, for quality loss and failure detection, images of machineries within the factory have to be labelled if it shows defects or not [35]. The labels can be e.g., defined manually (e.g., by crowdsourcing) or based on existing labels used within a [35]. In general, data labelling generates a huge effort for processing unstructured data within a data analytics project [28]. One of our experts clearly stated that a "efficiency in labelling the data" is needed not to waste too much time.

*Data Quality:*  Data quality problems often arise, for example, due to missing accuracy, completeness, consistency or time-relation issues [31]. Without suitable data quality, analysis based on this will may not result in valid findings. According to the assessment of our experts, data quality problems often occur with data provided by manufacturing (e.g., at the shopfloor level). One expert described these aspects in an example focusing on the gathering of voice data:

> *"Language often first has to be translated (automatically) into text. Here there are sometimes already the first quality problems, which then continue".*

Another expert also takes up this issue and adds the following:

> *"[...] consistency, completeness, unmatched timestamps".*

From these statements, it can be summarized that often the unstructured data that is available cannot be used due to poor quality. For this reason, it is an important aspect to address data quality. Only valid data can be used with the help of adequate analysis capabilities to carry out beneficial projects.

*Data Integration:* Particularly in an industry 4.0 setting, data is coming from different data sources (e.g., different manufacturing stations and produced IoT devices) with different data formats and have to be integrated into the analytical system [16, 19]. This diversity becomes even greater the more unstructured data there is, as this per se is more difficult to standardize. For this reason, data integration from shopfloor and different other sources is another critical path for using unstructured data in analytics. This can be illustrated by the insight provided by one of the experts:

*"No standardized ETL pipeline - great heterogeneity in the structure of log files, text files, etc. [...] Data integration is always an individual project."*

Data integration in itself is not only complex, but also very time-consuming. This means that companies that want to use the analysis of unstructured data must provide own capacities for this purpose. This is underlined by the following expert statement, which stand totally in line with the expert cited above:

*"[...] Data integration is [...] an individual project".*

It can be stated that especially in the area of unstructured data, data integration should be given special attention. If unstructured data are to be included in the analysis, then the integration effort increases considerably due to their particularly diverse structure. These must be taken into account before the project begins so that capacities can be planned, and measures defined.

## 4.3  Variety

Variety is another factor that our experts see as a critical challenge based on their experience. In general, the variety of data sources within smart factories is growing [21]. It follows that different methods must be used for the joint analysis of different data. Due to the different structure of data, different complex approaches like text mining, image recognition or audio analytics have to be applied and adjusted to the current data sources. One of our experts highlighted this challenge within the following statement and pointed out, that particularly unstructured data is affected by variety aspects:

*"The higher the percentage of unstructured data, the more complex the analysis of the database."*

As a consequence of the use of different analysis methods, the results are not uniform. For this reason, they cannot simply be interpreted together in a standardized way. The experts are also aware of this lack of harmonization, as the following statement shows.

*"Interpretation of results often incorrect - lack of standardization of procedures."*

The conclusion that can be drawn from this is that the integration of unstructured data into the analysis is always associated with a trade-off. If unstructured data is integrated, this leads on the one hand to a gain in information. However, the more unstructured data is integrated, the greater the variety. Of course, more information is then available by definition. But simultaneously, the number of analytical methods that have to be used increases. This not only favors the occurrence of errors in the execution and combination of the individual analyses, but in particular also the probability of misinterpretations in the findings. Hence, it is important to find the right balance here.

## 4.4 Measures

Additionally, the experts also reported about the measures currently taken in practice to use unstructured data for analytics in the industry 4.0 context. At the moment, they are mostly focusing on running "lighthouse" projects and investing in trainings for the employees as stated by the experts like following:

> *"Teams of data analysts are being built; lighthouse projects are being initiated"*
> *"Investing in training for DevOps and data engineering"*

However, there are still numerous open points that should be improved through targeted measures. As a main aspect here, experts mentioned an improved management support in awareness of occurring challenges, providing resources, the implementation of agile organizational structure as well as improvements in infrastructure and software tools.

Summarizing the current measures, it is evident that they should be more structured and contextualized. Building on the experts' assessments, three fundamental challenges exist that are of great importance when unstructured data is to be included in the analysis. These could be used as an initial starting point to derive concrete measures. In example, looking at the initial framework of this research helps the management to understand the challenges. Hence, the awareness increases and especially managerial decision like an agile organization structure can be triggered. In sum, that leads to an improvement and increases the probability of running successful projects.

## 5   Conclusion, Discussion and Future Research

Although more and more unstructured data is available through more and more connected devices and sensors [21] enterprises are reluctant to use the data, particularly in the context of industry 4.0. From the holistic view of the challenges presented in the framework, it can be deduced that companies may fail because concepts for the analysis of structured data cannot be transferred directly. The results of the study address this issue. The presented study revealed three major challenges occurring when unstructured data is integrated into data analysis in the industry 4.0 context. Based on the results a framework of data analytics challenges of unstructured data for industry 4.0 was developed.

The findings contribute to the current body of knowledge in research and practice. Our research identified a lack of analytical know-how within the organization, several production data issues including data labeling, data quality and data integration as well as increased analytical variety. It highlights the importance for addressing these challenges to run successful data analytics projects and extends current knowledge of applying analytical approaches with unstructured data. The proposed framework can be used to understand why the outcome of the usage of unstructured data is different to the usage of only structured data for analytical projects. The usage of the framework can deliver different viewpoints and action areas for an organization regarding e.g., analytics, integration, and quality aspects. From the scientific view point the findings extend research in several directions. We advanced the definition of data model-related requirements of Gölzer et al. [19] such as the unification of semantics and data integration for unstructured data. Furthermore, our work drives the development of requirements and architectural frameworks for industry 4.0 such as recommended by Gokalp et al. [18]. We also extended the research of Li et al. [6] who identified the integration of management systems as an important enabler for big data in industry 4.0. In practice, managers can use our results improve their understanding of running analytics project under the use of unstructured data in the industry 4.0 context. The resulting awareness can help to avoid failing projects. Hence, recommendations for targeted measures and an adjusted project setup can be derived from our findings. We pre-liminary identified first possibilities like ideas of learning within lighthouse and pilot projects as well as better trainings and management support and related awareness.

No research is without limitations. First, the interviews were conducted using a semi-structured questionnaire send to the participants informal via email. Although, the possibility to contact the researchers was given none of the participants started a discussion or asked a follow-up question. Therefore, it might be enriching to conduct face-to face interviews with the same questionnaire again to strengthen-up the findings. Additionally, the sample itself can be mentioned as a limitation. We only asked leading experts from Europe. This picture may vary in other countries (like e.g., BRIC states) with different industrial focus and segment [37]. Therefore, we are planning to extend our current study to other countries for a comparison and validation of the results. Furthermore, how strong the challenges vary and influence the analytical project targets is still open. Therefore, different studies (quantitative and qualitative) have to be setup for further investigations. A detailed examination how organizations can capture these identified challenges is missing and is currently under research. Some identified challenges (e.g., aspects of data issues) may also occur in other industry sectors and should be further explored. Besides the limitations, which can be taken up in the future research the initial framework highlights opportunities for future research in this collaborative research project. A next step on the research agenda might be to deeper investigate the data integration. In example, it might be very interesting to investigate how different data sources should be evaluated due to their meaning of their provided information. Therefore, a deeper knowledge about available data and their importance in different parts of industry 4.0 must be examined.

# References

1. Provost, F., Fawcett, T.: Data Science for Business: What You Need to Know about Data Mining and Data-analytic Thinking. O'Reilly Media, Sebastopol (2013)
2. Harbart, T.: Tapping the Power of Unstructured Data. MIT Sloan Management School (2021)
3. Davenport, T., Guszcza, J., Smith, T., Stiller, B.: Analytics and AI-driven enterprises thrive in the Age of With. Deloitte Insights (2021)
4. Bordeleau, F.-E., Mosconi, E., Santa-Eulalia, L.A.: Business intelligence in industry 4.0: state of the art and research opportunities. In: Proceedings of the 51st Hawaii International Conference on System Sciences (HICSS), Waikoloa, HI, USA (2018)
5. Lasi, H., Fettke, P., Kemper, H.-G., Feld, T., Hoffmann, M.: Industry 4.0. Bus. Inf. Syst. Eng. **6**(4), 239–242 (2014)
6. Li, G., Tan, J., Chaudhry, S.S.: Industry 4.0 and big data innovations. Enterp. Inf. Syst. **13**(2), 145–147 (2019)
7. Li, S., Xing, F., Peng, G., Liang, T.: Enablers for embedding big data solutions in smart factories: an empirical investigation. In: PACIS 2019 Proceedings, X'ian, China (2019)
8. Kähkönen, T., Alanne, A., Pekkola, S., Smolander, K.: Explaining the challenges in ERP development networks with triggers, root causes, and consequences. Commun. Assoc. Inf. Syst. **40**(1), 249–276 (2017)
9. Winter, S.J., Butler, B.S.: Creating bigger problems: grand challenges as boundary objects and the legitimacy of the information systems field. J. Inf. Technol. **26**(2), 99–108 (2011)
10. Webster, J., Watson, R.T.: Analyzing the past to prepare for the future. MIS Q. **26**(2), 494–508 (2002)
11. Al-Abassi, A., Karimipour, H., HaddadPajouh, H., Dehghantanha, A., Parizi, R.M.: Industrial big data analytics: challenges and opportunities. In: Choo, K.-K., Dehghantanha, A. (eds.) Handbook of Big Data Privacy, pp. 37–61. Springer, Cham (2020). https://doi.org/10.1007/978-3-030-38557-6_3
12. Alcácer, V., Cruz-Machado, V.: Scanning the industry 4.0: a literature review on technologies for manufacturing systems. Eng. Sci. Technol. **22**(3), 899–919 (2019)
13. Arnold, L., Jöhnk, J., Vogt, F., Urbach, N.: A taxonomy of industrial IoT platforms' architectural features. In: Ahlemann, F., Schütte, R., Stieglitz, S. (eds.) WI 2021. LNISO, vol. 48, pp. 404–421. Springer, Cham (2021). https://doi.org/10.1007/978-3-030-86800-0_28
14. Baiyere, A., Topi, H., Venkatesh, V., Wyatt, J., Donnellan, B.: The internet of things (IoT): a research agenda for information systems. Commun. Assoc. Inf. Syst. **47**(1), 557–579 (2020)
15. Baran, M.L., Jones, J.E.: Mixed Methods Research for Improved Scientific Study. IGI Global, Hershey (2016)
16. Dong, X.L., Halevy, A., Yu, C.: Data integration with uncertainty. VLDB J. **18**(2), 469–550 (2009)
17. Fay, M., Kazantsev, N.: When smart gets smarter: how big data analytics creates business value in smart manufacturing. In: ICIS Proceedings, San Francisco (2018)
18. Gokalp, M.O., Kayabay, K., Akyol, M.A., Eren, P.E., Koçyiğit, A.: Big data for industry 4.0: a conceptual framework. In: International Conference on Computational Science and Computational Intelligence (CSCI), Las Vegas, pp. 431–434 (2016)
19. Gölzer, P., Cato, P., Amberg, M.: Data processing requirements of industry 4.0-use cases for big data applications. In: ECIS 2015, Münster (2015)
20. Gröger, C.: Building an industry 4.0 analytics platform. Datenbank-Spektrum **18**(1), 5–14 (2018)
21. Jasperneite, J., Sauter, T., Wollschlaeger, M.: Why we need automation models: handling complexity in industry 4.0 and the internet of things. IEEE Ind. Electron. Mag. **14**(1), 29–40 (2020)

22. Khanra, S., Dhir, A., Mäntymäki, M.: Big data analytics and enterprises: a bibliometric synthesis of the literature. Enterp. Inf. Syst. **14**(6), 37–768 (2020)
23. Köhler, M.: Industry 4.0: Predictive maintenance use cases in detail. Bosch ConnectedWorld Blog (2018)
24. Matavire, R., Brown, I.: Profiling grounded theory approaches in information systems research. Eur. J. Inf. Syst. **22**(1), 119–129 (2013)
25. Müller, O., Junglas, I., Debortoli, S., vom Brocke, J.: Using text analytics to derive customer service management benefits from unstructured data. MIS Q. Exec. **15**(4), 243–258 (2016)
26. Myers, M.D., Avison, D.E.: Qualitative Research in Information Systems: A Reader. SAGE, London (2002)
27. Pauli, T., Lin, Y.: The generativity of industrial IoT platforms: beyond predictive maintenance? In: ICIS 2019, Munich (2019)
28. Roh, Y., Heo, G., Whang, S.E.: A survey on data collection for machine learning: a big data-AI integration perspective. IEEE Trans. Knowl. Data Eng. **33**(4), 1328–1347 (2019)
29. Sahal, R., Breslin, J.G., Ali, M.I.: Big data and stream processing platforms for Industry 4.0 requirements mapping for a predictive maintenance use case. J. Manuf. Syst. **54**, 138–151 (2020)
30. Salvadorinho, J., Teixeira, L.: Shop floor data in Industry 4.0: study and design of a manufacturing execution system. In: CAPSI 2020 Proceedings (2020)
31. Scannapieco, M., Missier, P., Batini, C.: Data quality at a glance. Datenbank-Spektrum **14**(1), 6–14 (2005)
32. Möhring, M., Schmidt, R., Keller, B., Sandkuhl, K., Zimmermann, A.: Predictive maintenance information systems: the underlying conditions and technological aspects. Int. J. Enterp. Inf. Syst. **16**(2), 22–37 (2020)
33. Tao, F., Qi, Q., Liu, A., Kusiak, A.: Data-driven smart manufacturing. J. Manuf. Syst. **48**, 157–169 (2018)
34. Wang, J., Ma, Y., Zhang, L., Gao, R.X., Wu, D.: Deep learning for smart manufacturing: methods and applications. J. Manuf. Syst. **48**, 144–156 (2018)
35. Whang, S.E., Lee, J.-G.: Data collection and quality challenges for deep learning. VLDB Endow. **13**(12), 3429–3432 (2020)
36. Xu, L.D., Duan, L.: Big data for cyber physical systems in industry 4.0: a survey. Enterp. Inf. Syst. **13**(2), 148–169 (2019)
37. Schmidt, R., Möhring, M., Härting, R.-C., Reichstein, C., Neumaier, P., Jozinović, P.: Industry 4.0 - potentials for creating smart products: empirical research results. In: Abramowicz, W. (ed.) BIS 2015. LNBIP, vol. 208, pp. 16–27. Springer, Cham (2015). https://doi.org/10.1007/978-3-319-19027-3_2
38. Lee, J., Ardakani, H.D., Yang, S., Bagheri, B.: Industrial big data analytics and cyber-physical systems for future maintenance & service innovation. Procedia CIRP **38**, 3–7 (2015)
39. Recker, J.: Scientific Research in Information Systems: A Beginner's Guide. Springer, Berlin (2013). https://doi.org/10.1007/978-3-642-30048-6

# Are We Speaking the Same Language?
# An Analysis of German and Chinese Local
# Shopping Platforms

Sören Aguirre Reid[1]([envelope]), Richard Lackes[1], Markus Siepermann[2],
and Valerie Wulfhorst[3]

[1] Technische Universität Dortmund, Dortmund, Germany
{soeren.aguirrereid,richard.lackes}@tu-dortmund.de
[2] Technische Hochschule Mittelhessen, Giessen, Germany
markus.siepermann@mni-thm.de
[3] Fachhochschule Südwestfalen, Soest, Germany
wulfhorst.valerie@fh-swf.de

**Abstract.** Local owner-operated retail outlets (LOOROs) are struggling world-wide due to increasing online competition with e-marketplaces and changing customer behavior. In this competitive environment, Local Shopping Platforms (LSPs) seem to be a promising vehicle to help LOOROs overcome their manifold digitalization difficulties. The development of LSPs in China shows that they convince LOOROs to join them and attract customers through their location-dependent service offers. In contrast, German LSPs struggle to utilize locational advantages, and LOOROs are discouraged from joining them. Hence, this paper examines existing Chinese LSP and compares them with German LSPs. Moreover, this study uses Hofstede's cultural dimension as a theoretical lens to investigate if the offered location-dependent services among German and Chinese are culturally driven. The results revealed that Chinese LSPs better understand how to provide services that integrate the store as a customer touchpoint and strengthen the locational advantage of LOOROs. Regarding Hofstede's cultural dimension, uncertainty avoidance is a cultural barrier for German LSPs and LOORO to try new digital services compared to their more adaptable Chinese counterparts.

**Keywords:** Local Shopping Platforms · Location-based services ·
Location-enabled services · Hofstede cultural dimensions

## 1 Introduction

Local owner-operated retail outlets (LOOROs), like traditionally stationary retailers or other small-and medium-sized enterprises (SME) are struggling worldwide due to increasing online competition with e-marketplaces and the changing customer behavior towards new services and multichannel shopping behavior [1–4]. Therefore retail outlets or LOOROs can be characterized by a small-sized store area, specific products (e.g., fashion, toys, jewellery), a restricted number of employees and a high degree

Ē. Nazaruka et al. (Eds.): BIR 2022, LNBIP 462, pp. 126–140, 2022.
https://doi.org/10.1007/978-3-031-16947-2_9

of owner-involvement in the business operations [3]. The intense business and market transformation lead to many stores closing in city centres. For instance, up to 64,000 local retailers in Germany are expected to be at risk before 2030, without considering the coronavirus's impact [2]. Reasons for the struggle of German LOOROs are manifold. For instance, LOOROs face internal barriers (e.g., lack of digital knowledge, financial capital, and time) and external barriers (e.g., lack of training options, lack of support), which prevent LOOROs from transforming their business models [3–5]. Like German LOOROs, Chinese LOOROs have also faced a strong sales decrease, with store closures increasing due to changing customer behavior and rapidly-growing online retail sales since 2014 [6]. Moreover, Chinese LOOROs also lack resources (e.g., lack of online presence), and their online stores have not been successful [7–9]. However, the situation has changed since 2017. Chinese city centres have recorded a noticeable shift among young customers from online back to traditional shopping in physical stores, shopping malls and mono-brand retail stores. Even the Covid-19 pandemic has not changed this trend [9]. One of the main drivers behind the comeback of Chinese LOOROs are local shopping platforms (LSP). For instance, LOOROs accounted for 89% of all retailers on the Chinese LSP Meituan [9].

So-called local shopping platforms (LSP) act as intermediaries and inter-organizational service hubs between LOOROs and their customers. Unlike classic e-marketplaces, LSPs focus on customers and/or retailers from a certain area. LSPs use this criterion as their unique selling proposition and try to utilize the locational advantages of LOOROs by offering location-dependent services to navigate the customer to the store [10, 11]. In the case of China, LSPs enable the integration of online and offline touchpoints that increase the customer in-store experience [12, 13]. These integrated touchpoints enable digital services like in-store self-check-out, the possibility of checking offline stock and personalized delivery options (e.g., same-day delivery). The Chinese case highlights that the location of LOOROs *"is not dead"* and is an essential aspect of the rising service competition between pure online players and LOOROs [14]. In contrast to the development in China, former studies criticize that German LSPs do not help LOOROs attract more customers to their stores, that they do not utilize the locational advantages of LOOROs as a unique selling proposition, and that they do not help generate higher revenues [11, 15, 16]. Even during the Covid-19 pandemic, LOOROs have still been reluctant to join a platform and then leave as soon as the easing of the situation comes in sight [4, 17, 18]. Therefore, German LSPs providers need to understand their Chinese counterparts' approaches and offered services, which may can help provide more appropriate services for LOOROs and their customers.

For the study, we chose two countries representing two extremes regarding their digitalization level of the retail market: China is setting the present and future standards for retail. Chinese companies are pioneering in developing, using and providing new digital technologies and services. [19, 20] Of course, new digital technologies and services also have adverse effects like low customer data protection and the need for caution [18]. In contrast, German retailers or LSPs hesitate to implement new technologies and services or underestimate their customers' needs [4]. Moreover, developments in the Chinese retail market also affect the scientific literature. For instance, the development of live-stream shopping in China has led to a new research stream and has highly relevant practical insights into the slow diffusion of live-stream shopping in European retail markets [21]. Therefore, the first and second research questions analyze the existing LSPs approaches and offer location-dependent services in both countries.

*RQ1: How do existing Local Shopping Platform approaches differ among German and Chinese platforms?*

*RQ2: How do existing Local Shopping Platforms offer location-dependent services differ among German and Chinese Local Shopping Platforms?*

However, it is not enough to descriptive analyze the offered services of LSPs in both countries. Since China has become the pioneer in setting the standards for the present and future of retail, foreign marketplaces (e.g., Germany, USA) tend to directly copy ideas from those marketplaces without considering the cultural differences. However, it has been proven that the ignorance of cultural differences will cause the failure of those services or systems [22]. Because the culture impacts the ways, local customers evaluate and use services [23]. Hence, this study aims to shed some light on how the underlying culture shapes the service provision of LSPs. Despite the importance of culture for the service provision, none of the former studies investigated how the countries' culture affects the service provision of platforms in different geographical areas in general [23, 24]. For the analysis, the current study uses Hofstede's cultural dimension [25, 26] as a theoretical lens to explain the offered location-dependent service by Chinese and German LSPs (see RQ3). The Hofstede's cultural dimensions (1980) [25, 26] is the most used approach for culture comparisons [27]. Moreover, studies from various contexts use Hofstede to explain cultural differences regarding using services or apps like mobile banking/shopping [28–30].

*RQ3: Can Hofstede's cultural dimension explain the offer of location-dependent services among German and Chinese Local Shopping Platforms?*

Research on LSPs has many ties to research on e-marketplaces and e-intermediaries (see subchapter Background: Local Shopping Platforms). First, this study contributes new theoretical and practical insights to e-marketplace research in general. It contributes to the little-explored area of LSPs, a subset of e-marketplaces. Second, none of the former studies investigated how culture affects the service provision of platforms in different geographical areas in the field of LSPs. Former studies have investigated LSPs in general [15], focusing on participating LOOROs [16, 31] or the LSP customers [5, 31], and analyzing the existing types and services offered by LSPs [10, 11]. Third, former research regarding the cultural context only investigated the customer perspective and not the service provider perspective [28–30].

This paper follows a 2019 sample of a preliminary study using content analysis [11]. To compare the results of both countries, we collected the data for the Chinese LSPs with a content analysis and compared the results with the preliminary study [11]. The paper is structured as follows: In Sect. 2, we discuss the existing literature and the theoretical background for our study. In Sect. 3, we introduce the methodological foundation and present each step of the content analysis for the data. In Sect. 4, we present our results. Section 5 concludes, identifies limitations and discusses future research opportunities.

## 2 Background: Local Shopping Platforms

An LSP for LOOROs takes the form of a multi-sided platform which mediates different groups of users, such as buyers and retailers [32]. As two-sided Markets, LSPs match buyers and sellers in a relationship where the value for one group increases as the number of participants from the other group increases [33]. LSPs function as an intermediary, enabling direct interaction between the groups of users, facilitating the exchange of information [34], goods and services, and transaction and fulfillment services [35]. Furthermore, LSPs limit themselves (self-restriction criteria), focusing on customers or/and retailers from a certain area. The self-restriction criterion meets the needs of LOOROs, which consider the targeting of customers who are too far away to be a waste of their marketing budget [15]. The regional focus distinguishes LSPs from e-marketplaces and e-intermediaries such as Amazon and Alibaba, where the boundaries between Business-to-Business (B2B) and Business-to-Consumer (B2C) blur, as well as regional and national restrictions [36]. In line with local restrictions, the research considers LSPs to be geographically restricted and target customers living in a specific region or city [31]. Bärsch et al. 2019 specify this local component and introduce a self-restriction criterion as an identifier of LSPs. The self-restriction criterion considers if an LSP cooperates only with retailers from a certain area, just doing business with customers from a certain area, or both at the same time [10]. This contrasts with the specific location focus. Schade et al. (2018) interpret LSPs as a response to the adverse outcomes (e.g., loss in total revenue) of the e-commerce trend [15].

### 2.1 Types of Shopping Platforms

Previous studies show that LSPs provide a non-standardized and diverse service landscape, like e-marketplaces [11, 37]. The services offered range from store opening hours and product information to a full transaction process with pricing and logistics. LSPs can also be distinguished by the level of collaboration between the participants and the necessary front end information and communication technology (ICT) investment [38]. According to this classification, our LSPs require a large amount of collaboration, preferably with low ICT investment. In line with prior research, we will use a typological categorization of LSPs based on their e-marketplace functionalities and local focus [10, 37]. The first function (match of buyers and sellers) allows for the identification of two categories: Store Locator Platforms and Product Catalogue Platforms. Store Locator Platforms offer rudimentary information about opening hours, whereas Product Catalogue Platforms provide an overview of product information (e.g., size, color). From the second function (exchange of information), an additional platform category (Product Enquiry Platforms) can be derived. Product Enquiry Platforms enable customers to request additional product information from the seller and check the product's availability in-store. The third function (transaction and fulfillment) of e-marketplaces helps introduce another two categories: Affiliate Transaction and Full Transaction Platforms. Affiliate Transaction Platforms allow for the purchase of products, but customers complete the transaction process on an external website with the help of an affiliate shop. Full Transaction Platforms offer the full e-marketplace service range, all the way up to complete transactions, payment and logistic services [10, 11].

## 2.2   Categories of Services on Local Shopping Platforms

Historically, location has been the most crucial decision (even for successful retailers) because a poor location is an insurmountable obstacle [39]. The omnipresence of digital platforms in society and businesses - particularly in e-commerce - is challenging high street retail locations [14, 40]. Prime locations in attractive, large cities are expected to suffer minimally when compared to smaller cities. Moreover, digital transformation has bridged the spatial distance with internet connections, smart devices and digital services and increased the pressure of the location on LOOROs in city centers [40]. However, even in an environment with near-zero trade costs, physical distance matters due to the negative effects of growing transportation costs on online sales prices and customer demand. Therefore, relying on location without connecting to appropriate services will not help LOOROs in the rising service competition with pure e-commerce players, who also offer location-dependent services [41]. Empirical research provides evidence for the importance of location-dependent services for the customers' intention to purchase and also shows that customers enjoy using location-dependent services (e.g., same-day delivery) [5, 42]. Moreover, location-dependent services attract and retain customers in multi-channel retail environments [14, 43].

LSPs make use of locational advantages, and two LSP service types can be identified: location-independent and location-dependent services [10]. Location-independent services are generally available services that are not bound to a specific customer position. They include standard web services, such as payment services and marketing channels (e.g., Instagram, TikTok). Location-dependent services instead make use of the position of customers. They can be further divided into location-based and location-enabled services. Location-enabled services are feasible if the retailer's location is close to the customer's household. This allows for short-distance services with low transportation costs, such as same-hour delivery or clicks and return services. Location-based services are feasible if the customer is close to the store location. Location-based services aim to utilize foot traffic in popular places (e.g., main streets) and use customer devices (e.g., smartphones) as location-awareness information systems [10, 44]. Location-based services consider location-based navigation, marketing, consulting, discounts and self-checkout services.

## 2.3   Hofstede's Cultural Dimensions

In line with Hofstede, we consider culture as the result of outside influences (e.g., new technology) that affect ecological factors (e.g., technology, economics), which lead to a gradual adjustment in social norms and trigger behavior as a consequence [25]. Therefore we consider the cultural dimension as a proxy of culture and follow Hofstede's classification along four dimensions: power distance (PDI), uncertainty avoidance (UAI), individualism-collectivism (IDV), and masculinity-femininity (MAS). Later, a fifth and sixth dimension were added: the long/short-term orientation and indulgence. However, the scope of Hofstede's cultural dimensions is not without limitations and cannot fully explain the national culture of Germany or China. Therefore, we solely focus on LSPs service provision and neglect the existing types and customer perspectives, which is a limitation of this study. For the existing types, we only knew the current LSP type of a provider but did not know the reason for the decision (e.g., legal system, taxation, and so). For further analysis, we do not consider the cultural dimensions of masculinity versus femininity (MAS) (which "[…] indicates that the society will be driven by competition, achievement and success, […] the dominant values in society are caring for others and quality of life" [45]), long term orientation versus short term normative orientation (LTO) (indicating "[…] how every society has to maintain some links with its own past while dealing with the challenges of the present and future […]" [45]) and indulgence versus restraint (IVR) (depicting "[…] the extent to which people try to control their desires and impulses[…]" [45]) because the Chinese and German culture do not vary along those dimensions. Based on the estimation values of the cultural dimension by Hofstede, the Chinese and German culture can be characterized along the three dimensions as follow (see Table 1): Germany has a low PDI value, indicating that inequality amongst people is not acceptable. Moreover, direct and participative communication is expected. Authorities' control is disliked.

According to Hofstede's estimated values, the PDI in China is high. The high value in China reflects that inequalities amongst people are acceptable, and individuals are influenced by formal authority [26, 46]. The value of PDI in China fits together with the individuals' role in a group and reinforces collectivism [47]. In contrast to Germanies' direct and participative communication, Chinese people avoid direct confrontation and circle around problems. Otherwise, a confrontation would stress harmony and the need to consider the interests and understanding of the other party [48]. The value of IDV shows that China is a highly collectivist culture where people act in the group's interests and not necessarily of themselves [26, 46]. For instance, in China, requests to one person are passed on to other persons who are entitled to take the decision [47]. Another example, social interaction is perceived in terms of collectivism and social usefulness. In contrast, western countries like Germany consider social interaction in light of competitiveness, self-confidence, and freedom [49, 50]. This is in line with the strong belief in the ideal of self-actualization for individualist cultures like Germany [26, 46].

The score for UAI reveals that the Chinese culture can be characterized as comfortable with ambiguity (e.g., the Chinese language is full of ambiguous meanings) [26, 45, 46]. Moreover, the Chinese are adaptable and entrepreneurial, which is a clear advantage concerning digital transformation. Therefore, it is not surprising that Chinese companies are pioneering in providing digital services, and customers are more willing to embrace new technology services [20]. In contrast, there is a slight preference for uncertainty avoidance in Germany. For instance, this is reflected by the law system or the strong preference for planning and well-thought-out projects in Germany [4, 26, 46]. But also German customers feel threatened by uncertain or ambiguous situations like new technology services [51].

# 3 Methodology

LSPs have an unstandardized, multiple-media-mixed website structure (e.g., graphics or different names for their services). In order to address this issue, we will follow former studies with similar challenges and conduct a content analysis [10, 52, 53]. We will undertake the content analysis in line with the guidelines set by Krippendorff [54]. After

**Table 1.** Cultural dimension and country comparison Germany (DE) & China (CN)

| Cultural dimension | Definition cultural dimensions | Values* | |
|---|---|---|---|
| | | DE | CN |
| PDI | "[…] the extent to which the less powerful members of institutions and organizations within a country expect and accept that power is distributed unequally." [45] | 35 | 80 |
| IDV | "[…] the degree of interdependence a society maintains among its members." [45] | 67 | 20 |
| UAI | "The extent to which the members of a culture feel threatened by ambiguous or unknown situations and have created beliefs and institutions that try to avoid these […]". [45] | 65 | 30 |

*Estimated values of the cultural dimension based on Hofstede [45].

defining the research scope and questions, we identified the existing LSPs through an explorative web search. In the second step, we conducted a pre-test with the codebook and 30% of the identified LSPs in China. A revision procedure followed this pre-test to improve the categorization and streamline the coding. Subsequently, three individual Chinese coders conducted the content analysis in a third step. In the fourth step, we calculated the intercoder reliability with two reliability measures used frequently in existing research [55]. Finally, an expert panel of senior researchers discussed discrepancies in the coding results and resolved them. The complete research steps will be discussed in detail in the following.

### 3.1 Sample, Pre-test and Coding of the Content Analysis

To get the LSPs, we will use a multiple-stage process with the relevant sampling approach and self-restriction criterion [54]. For the development of the keyword list, we will consider previous studies in the field and narrow the keyword list to the local city name because some LSPs name themselves after their city [15]. For the web search, we will use the keyword list of the preliminary study [11]. Admittedly, we translated the keyword list to Chinese and then back to German to ensure translation equivalence. The translation was conducted independently by two Chinese-speaking research assistants [56]. The online search process took place in June 2020 and was conducted on the Chinese website Baidu with the same keyword list from the preliminary study in Chinese [11]: LOORO + Local + (E-Marketplace, Shopping Platforms, Shops Online, Vendors Online, Marketplace, Products Online, Retail Online, Online Shop, Retailer Archive, Product Archive, Product Enquiry), Buy Local and City Name. To address the superficiality of the keyword list, we used the self-restriction criterion to improve the quality of the findings with a selection and screening process. We found 66 LSPs, and we excluded 46 platforms for several reasons, e.g., the website was inactive, because it was a business-to-business platform, or because it did not fit the selection criterion. We reduced the first set to a final set of 20 Chinese LSPs (e.g., Nuo Ya (喏呀), Lu Xi Tong Cheng (泸溪同城) or E Le Me (饿了么)). Our preliminary study identified 102 LSPs in Germany (e.g., Atalanda, like Lippstadt, Hallo Altmark) [11]. In the second step, we adapted the codebook of extant literature with a focus on LSPs to improve the reproducibility of the content analysis [10, 54]. Based on the codebook, we defined coding units to distinguish between the separate descriptions assigned to the categories (location-based, location-enabled or location-independent services) [54]. We then conducted a pre-test, categorizing 30%, of the newly-identified LSPs and their services for China. The pre-test is necessary for ensuring consistent coding. The pre-test revealed 71 possible items (coding units): 5 typological items and 66 service items for the Chinese LSPs codebook. Based on the pre-test in China, we adjusted the payment options and marketing channels for China (e.g. added WeChat Pay). In the fourth step, each coder received a written introduction in Chinese to the general coding process (content analysis), which explained each typology and service item with an example to address any uncertainties [55, 57]. Each typological and service item was considered with the following rule for the codebook: "1" if Yes/Available on the website/mobile app, and "0" if No/Not Available. The typological score ranged from 0 to 5, and the service score ranged from 0 to 66. Concerning the typological score, the highest function defines the platform type. For instance, the Mei Tuan (美团) LSP fulfils the first and second functions and the third function. Therefore, we consider this LSP as a full transaction platform.

### 3.2 Intercoder Reliability of the Content Analysis

In the fourth step of the procedure, we calculated the level of agreement between the coders' decision with two different intercoder reliabilities [58]. The Holsti coefficent take values of .00 (no agreement) to 1.00 (perfect agreement). However, this approach tends to overinflate the result of the agreement. This is why we also consider Krippendorff's $\alpha$ as a measurement for intercoder reliability [57]. Krippendorff's $\alpha$ defines two reliability scale points as 1.000 for perfect reliability and 0.000 for the absence of reliability [55]. In line with Hayes and Krippendorff [55], we use a macro to compute the Krippendorff's $\alpha$. Regarding the five typological items, each coder had to judge 100 typology items for China (20 LSPs*5 typological items). For the typological items, the coders for China (Holsti: 0.98) achieved a very good intercoder reliability [59]. Concerning Krippendorff's $\alpha$, the Chinese coder ($\alpha$: 0.95) achieved a very good level [54, 55]. Regarding the service items, each coder had to judge 1320 service items for China (20 LSPs* 66 service items). For the service items, the Chinese coders (Holsti: 0.92) achieved a very good level of intercoder reliability [59]. Concerning Krippendorff's $\alpha$, the Chinese coders ($\alpha$: 0.833) achieved an acceptable degree of reliability [54, 55]. For the final step, an expert panel of four senior researchers with high expertise in the field of e-marketplaces (particularly LSPs) discussed the coder inconsistencies for each platform and harmonized all remaining discrepancies (638 items for China) for the final sample. In the next chapter, we compare the results of the primary study [11] with the results for the Chinese LSPs.

# 4   Results

## 4.1   Local Shopping Platforms Approaches in Germany and China

The primarily study revealed 65 Store Locator Platforms, three Product Catalogue Platforms, ten Product Enquiry Platforms, two Affiliate Transaction Platforms, and 22 Full Transaction Platforms in Germany [11]. For China, the content analysis revealed 20 Full Transaction Platforms. Unlike China, most LSPs in Germany offered only contact and store location information and did not provide the full e-marketplace service.

## 4.2   Local Shopping Platform Services in Germany and China

The platforms in both countries provide 29 location-dependent services (16 location-enabled and 13 location-based services) and 40 location-independent services in Germany and 37 in China. We can only compare platforms in both countries that fulfilled the third function (transaction and fulfillment), because platforms that fulfill the first and second functions of e-marketplaces do not exist in China.

The results of comparing Chinese and German LSPs indicated a higher level of location-enabled services for Chinese LSPs. Like German LSPs, Chinese LSPs provide a high level of information services (e.g., opening hours or contact information). However, Chinese LSPs have a higher offering of communication services (e.g., feedback and community options and loyalty cards). Moreover, all Chinese LSPs offer multiple fulfillment services (e.g. same-day delivery and click & collect). Only six German LSPs offered multiple fulfillment services for their customers, with a combination of same-day delivery and click & return/collect services. However, the majority offered only one delivery option. The location-based service level offered in China is significantly higher than in Germany. Chinese LSPs offer multiple information services (e.g., barcode scanner 15 LSPs; location-based maps 12 LSPs) and navigation services (e.g., shopping guides 11 LSPs; outdoor navigation 14 LSPs). However, only two LSPs offer location-based payment with a self-checkout function in China. Only three German LSPs offer location-based services with information services (location-based maps for the next store (1 LSPs) and navigation services (outdoor navigation (2 LSPs). None of the LSPs in either country offered location-based communication services. Both countries have a high number of location-independent services. LSPs in both countries offer multiple recommendation services but differ slightly regarding the logistic services (e.g., more product availability services online or delivery on demand in China). Concerning payment services, LSPs in both countries offer card-based and digital wallet payments. Chinese LSPs have a stronger focus on the payment via the digital wallet (e.g., WeChat Pay or Alipay), which contrasts with card-based payment options (EC or credit) in Germany. The Chinese LSP communication and support services are only driven by social media (e.g., Weibo and WeChat). German LSPs also strongly focus on social media (e.g., Facebook and Instagram) and offer newsletters (14 LSPs) as a communication medium. Interestingly, 16 out of 20 Chinese LSPs provide their service solely via a mobile app (instead of a website). Only four LSPs offered their services via a website in China and had the lowest location-based service. Only five LSPs in Germany provided a mobile app.

# 5 Discussion

With regard to RQ1, the results confirmed the platform types derived from former studies [10, 11]. The study revealed that the Chinese LSP market can be characterized as full transaction platform-driven. This is in line with current business reports on the retail market [9, 18, 60]. The German LSP market can be characterized as information and communication driven due to the majority of store-locator, product catalogue and enquiry platforms. With regard to RQ2, the comparison of both countries revealed that Chinese LSPs offer a higher level of location-dependent services. They understand the importance of the mutual exchange of information and indirect communication between LOOROs and their customers, which contrasts with the strongly-criticized one-way communication approach of German LSPs [10, 15]. However, it is not only the communication that matters but also the fulfillment service. Chinese LSPs offered greater fulfillment services to address the last-mile problem of e-commerce [61]. Moreover, the higher logistic capabilities are in line with customer preferences for time reduction and increase the attractiveness of the e-marketplace [12, 43, 62], while German LSPs hesitate to extend their fulfillment service capabilities. Chinese LPS offer a higher level of location-based services to match their customers' needs [12, 15]. For instance, barcode scanners match the customer preference for multiple-channel shopping [9, 60, 63]. The results of the location-based services also confirm the importance of online-offline integration for customer adoption from the service perspective [12]. The results indicate that Chinese LSPs focus more on utilizing locational proximity between LOOROs and their customers. Apparently, they have understood that they depend on the existence of LOOROs and that only strong local partners can provide a sustainable basis for their platforms [31]. German LSPs still neglect the utilization of locational proximity with location-dependent services. The location-independent service findings show that Chinese LSPs utilize multiple social media accounts to address customer expectations [13]. German LSPs also utilize social media accounts but still use e-mail newsletters, which contrasts with customers' current social media usage [64].

For answering RQ3, we linked our findings with Hofstede's cultural dimensions. The results indicated that the cultural dimension of Hofstede can explain the service provision of LSP in Germany and China. In particular, China has a very high PDI value compared to Germany, which indicates that they are not using direct and open communication. Our results confirm the need for indirect communication with the high offering of indirect customer communication [26, 45, 46, 65]. Even the high number of chat options for customer support services for the location-independent services confirm this result. In the Chinese case, indirect communication help to avoid confrontation and maintain harmony with their customers or the LSPs providers [48]. An explanation is anonymity because LOOROs or customers may communicate more boldly than in a face-to-face situation (online disinhibition effect) [66]. Therefore, digital communication services help LOOROs and customers to overcome their cultural boundaries without losing their face in China. In line with the low PDI, LSPs and LOOROs or LOOROs with their customers have direct and open communication [26, 45, 46]. It is more critical for LOOROs to lure the customer into the store with location-enabled services like information services (e.g., maps with store locations, store opening hours, or store contact data) to have a face-to-face situation. Interestingly, according to the IDV dimension, location-enabled communication and support services (e.g., customer feedback and community integration) should be higher in individualism than in collectivist culture because less collective society tends to be uncertain even if products are selected as bestsellers. Thus, they need more reviews/customer feedback to judge the quality of the product [22]. Hence, the result of Germany does not confirm former studies [22, 67] and the cultural dimension. The low level of the UAI dimension in China is reflected by the higher offering of new digital services. For instance, the intense use of location-based services like information services (e.g., QR-code scanners), navigation services (e.g., location-based shopping tours), payment and billing services (e.g., self-checkout services) or location-enabled services like fulfilment services (e.g., click & return/collect), which confirms the Chinese characteristics of LSPs as adaptable and entrepreneurial [20, 45]. Chinese LSPs and the connected LOOROs seem more comfortable with new digital services than their German counterparts. A further explanation of the Chinese openness to new digital services is trust. In China, trust accumulates guanxi (network of relations) and mianzi (need to save face and self-protection). Therefore, LOOROs trust the quality of the provided services by the LSPs or related third parties like delivery services (e.g., same-day delivery) [47]. In the case of Germany, former studies show that LOOROs refuse to integrate their sales into an online marketplace or implement location-dependent services due to their internal and external barriers [4, 16]. Our study extends this finding by the cultural dimension. The cultural bond of uncertainty avoidance can also be seen as a factor which decoupled LOOROs from their near and far business environment [4]. Therefore, LOOROs are more concerned about the risks of LSPs or their offered services instead of the opportunities.

## 5.1 Practical Implications

The study provides several lessons for LSP Providers, their connected LOOROs and city manager. First, the service offerings of LSPs are cultural-driven. Therefore, LSP providers cannot easily copy offered services from foreign LSPs. In the first step, LSP needs to understand the cultural needs of their customers and then, in the second step, select the appropriate service to address them. This also holds for LOOROs to grow in their cultural environment (e.g., offering more cultural-related services like customer feedback options). Second, China highly offers indirect customer communication, which aligns with their cultural environment. In contrast, German LSPs' offering focuses on the information service needs and luring customers to the store. However, due to the increasing diffusion of smartphones, customer behavior is also changing in Germany, and the importance of indirect communication services (e.g., via chat options) is growing. The Chinese case shows that the store channel synergies with the mobile channel and increases purchase frequency and customer loyalty [53]. Both would satisfy German customers' direct communication needs and their changing preference for new communication channels (e.g., mobile apps). The existing approaches in China could provide a blueprint for German LSPs. Suppose LSPs, are still ignoring the changing customer behavior. In that case, customers will stop doing business with LSPs or the connected LOOROs after poor customer support experiences, which is the top reason for customers [68]. Third German LSPs and LOOROs are culturally bonded by their uncertainty avoidance in contrast to their adaptable and entrepreneurial Chinese counterparts. Local governments should hire "caretakers" to address this cultural barrier, who work together with LOOROs and support them. The low-threshold service by "caretakers" would fill an institutional gap. Moreover, a higher service provision of LSP positively influences customer attitudes and the intentions to buy on the LSP [5, 31]. For instance, the Chinese case shows that a higher logistic capability (e.g., more multiple fulfilment services offerings) would increase the attractiveness of the LSPs from a customer perspective [62].

### 5.2 Limitation and Future Research

To the best of our knowledge, this is the first study investigating the cultural impact on the service provision of LSP providers in two countries. Nevertheless, our study considers only three out of six dimensions of the Hofstede model. Second, the Hofstede model cannot explain further important factors like the legal system's impact on the chosen LSP type or the market environment (e.g., competition). Therefore, future research should investigate the cultural impact with various approaches like Porter's five forces. Third, our study identifies various services. However, we cannot analyze the actual customer use and fit regarding the cultural dimensions. Therefore, future research should analyze the actual customer behavior of location-dependent services and link them with the cultural dimensions. Fourth, we provide a written introduction to all coders in their language to address any uncertainties or misinterpretations, but we cannot rule out any biases for sure.

## References

1. Dholakia, U.M., Kahn, B.E., Reeves, R., Rindfleisch, A., Stewart, D., Taylor, E.: Consumer behavior in a multichannel, multimedia retailing environment. J. Interact. Mark. **24**, 86–95 (2010)
2. Köln, I.F.H.: Pressemitteilung Handelsszenario 2030: Wachstumsparadoxon im deutschen Einzelhandel. IFH Köln (2020)
3. Bollweg, L., Baersch, S., Lackes, R., Siepermann, M., Weber, P.: The digitalization of local owner-operated retail outlets: how environmental and organizational factors drive the use of digital tools and applications. In: 24th BIS, pp. 329–341. Hannover (2021)
4. Bollweg, L., Lackes, R., Siepermann, M., Weber, P.: Digitalization of local owner-operated retail outlets: between customer demand, competitive challenge and business persistence. In: 15th WI, pp. 1004–1018. Wirtschaftsinformatik, Potsdam (2020)
5. Bollweg, L., Lackes, R., Siepermann, M., Weber, P.: The role of location dependent services for the success of local shopping platforms. In: Abramowicz, W., Corchuelo, R. (eds.) Business Information Systems Workshops. LNBIP, vol. 373, pp. 366–377. Springer, Cham (2019). https://doi.org/10.1007/978-3-030-36691-9_31
6. Deloitte: China Retail Industry Development. Deloitte Global (2017)
7. Zhuang, G.: Structural change in China's retail industry in the first decade of the new century. J. Mark. Channels **20**, 288–324 (2013)
8. Sternquist, B., Huang, Y., Chen, Z.: Predicting market orientation: Chinese retailers in a transitional economy. Int. J. Retail Distrib. Manag. **38**, 360–378 (2010)

9. Fung Business Intelligence: What the Experts Say: Ten Highlights of China's Commercial Sector 2020. Fung Business Intelligence (2021)
10. Bärsch, S., Bollweg, L., Lackes, R., Siepermann, M., Weber, P., Wulfhorst, V.: Local shopping platforms – harnessing locational advantages for the digital transformation of local retail outlets: a content analysis. In: 14th WI, pp. 602–618. AIS, Gießen (2019)
11. Bärsch, S., Bollweg, L., Weber, P., Wittemund, T., Wulfhorst, V.: Local retail under fire: local shopping platforms revisited pre and during the corona crisis. In: Ahlemann, F., Schütte, R., Stieglitz, S. (eds.) Innovation Through Information Systems. LNISO, vol. 46, pp. 123–139. Springer, Cham (2021). https://doi.org/10.1007/978-3-030-86790-4_10
12. Yang, Y., Gong, Y., Yu, B., Zhang, J., He, T.: What Drives online-to-offline commerce: from a perspective of consumer. In: 15th WHICEB, pp. 289–296. AIS, Wuhan (2016)
13. McKinsey: China digital consumer trends in 2019. McKinsey & Company (2019)
14. Kim, T.Y., Dekker, R., Heij, C.: Cross-border electronic commerce: distance effects and express delivery in European union markets. Int. J. Electron. Commer. **21**, 184–218 (2017)
15. Schade, K., Hübscher, M., Korzer, T.: Smart retail in smart cities. best practice analysis of local online platforms. In: 15th ICETE, pp, 147–157. SCITEPRESS, Porto (2018)
16. Delgado-de Miguel, J.-F., Buil-López Menchero, T., Esteban-Navarro, M.-Á., García-Madurga, M.-Á.: Proximity trade and urban sustainability: small retailers' expectations towards local online marketplaces. Sustainability **11**, 1–20 (2019)
17. Appel, A., Hardaker, S.: Strategies in times of pandemic crisis - retailers and regional resilience in Würzburg Germany. Sustainability **13**, 1–19 (2021)
18. McKinsey: China consumer report 2021. Understanding Chinese Consumers: Growth Engine of the World. Special Edition. McKinsey & Company (2020)
19. Deloitte: The Path to Innovation Deloitte China Consumer & Industrial Products (2021)
20. Hardaker, S., Zhang, L.: Testing the water – prior-online market entry in China. Int. J. Retail Distrib. Manag. **49**, 1111–1129 (2021)
21. Abrahamsson, S., Harmath, B., Tolouee, A.: Live-Shopping in Europa: Das neue Must-have im Fashion- und Beauty-E-Commerce? Arvato Bertelsmann (2022)
22. Fang, H., Zhang, J., Bao, Y., Zhu, Q.: Towards effective online review systems in the Chinese context: a cross-cultural empirical study. Electron. Commer. Res. Appl. **12**, 208–220 (2013)
23. Zeithaml, V., Bitner, M., Gremler, D.: Services Marketing: Integrating Customer Focus Across the Firm. McGraw Hill, New York (2018)
24. de Reuver, M., Sørensen, C., Basole, R.C.: The digital platform: a research agenda. J. Inf. Technol. **33**, 124–135 (2018)
25. Hofstede, G.: Culture Consequences: International Difference in Work-Related Values. Sage Publications, London (1980)
26. Hofstede, G., Hofstede G. J., Minkov, M.: Cultures and organizations: Software of the mind. Revised and Expanded 3rd Edition. McGraw-Hill, New York (2010)
27. Annamoradnejad, I., Fazli, M., Habibi, J., Tavakoli, S.: Cross-cultural studies using social networks data. IEEE Trans. Comput. Social Syst. **6**, 627–636 (2019)
28. Chopdar, P.K., Korfiatis, N., Sivakumar, V.J., Lytras, M.D.: Mobile shopping apps adoption and perceived risks: a cross-country perspective utilizing the unified theory of acceptance and use of technology. Comput. Hum. Behav. **86**, 109–128 (2018)
29. Mortimer, G., Neale, L., Hasan, S.F., Dunphy, B.: Investigating the factors influencing the adoption of m-banking: a cross cultural study. Int. J. Bank Mark. **33**, 545–570 (2015)
30. Picoto, W.N., Pinto, I.: Cultural impact on mobile banking use – a multi-method approach. J. Bus. Res. **124**, 620–628 (2021)
31. Berendes, C.I., zur Heiden, P., Niemann, M., Hoffmeister, B., Becker, J.: Usage of local online platforms in retail: insights from retailers' expectations. In: 28th ECIS, pp. 1–11. AIS, Marrakech (2020)

32. Boudreau, K.J., Hagiu, A.: Platform rules: multi-sided platforms as regulators. In: Gawer, A. (ed.) Platforms, Markets and Innovation, pp. 163–191. Edward Elgar Publishing, Cheltenham (2009)

33. Eisenmann, T.R., Parker, G., van Alstyne, M.W.: Strategies for two sided markets. Harv. Bus. Rev. **84**, 92–101 (2006)

34. Armstrong, M.: Competition in two-sided markets. Rand J. Econ. **37**, 668–691 (2006)

35. Bakos, Y.: The emerging role of electronic marketplaces on the Internet. Commun. ACM **41**, 35–42 (1998)

36. Standing, S., Standing, C., Love, P.E.D.: A review of research on e-marketplaces 1997–2008. Decis. Support Syst. **49**, 41–51 (2010)

37. Petersen, K.J., Ogden, J.A., Carter, P.L.: B2B e-marketplaces: a typology by functionality. Int. J. Phys. Distrib. Logist. Manag. **37**, 4–18 (2007)

38. Smedlund, A.: Value cocreation in service platform business models. Serv. Sci. **4**, 79–88 (2012)

39. Achabal, D.D., Gorr, W.L., Mahajan, V.: MULTILOC: a multiple store location decision model. J. Retail. **58**, 5–25 (1982)

40. Becker, J., Betzing, J.H., Hoffen, M. von, Niemann, M.: A Tale of two cities. In: Riemer et al. (ed.) Collaboration in the Digital Age, pp. 291–307. Springer, Cham (2019)

41. Allen, J., et al.: Understanding the impact of e-commerce on last-mile light goods vehicle activity in urban areas: the case of London. Transp. Res. D: Transp. Environ. **61**, 325–338 (2018)

42. Ho, S.Y.: The effects of location personalization on individuals' intention to use mobile services. Decis. Support Syst. **53**, 802–812 (2012)

43. Lin, G., Chen, X., Liang, Y.: The location of retail stores and street centrality in Guangzhou. China. Appl. Geogr. **100**, 12–20 (2018)

44. Betzing, J.H: Beacon-based customer tracking across the high street: perspectives for location-based smart services in retail. In: 24th AMCIS. AIS, New Orleans (2018)

45. Hofstede-Insights. https://www.hofstede-insights.com/models/national-culture/. Accessed 27 July 2022

46. Hofstede-Insights. https://www.hofstede-insights.com/country-comparison/china,germany/. Accessed 27 July 2022

47. Moser, R., Migge, T., Lockstroem, M., Neumann, J.: Exploring Chinese cultural standards through the lens of German managers: a case study approach. IIMB Manag. Rev. **23**, 102–109 (2011)

48. HBR. https://store.hbr.org/product/lessons-from-abroad-when-culture-affects-negotiating-style/N0501B. Accessed 10 May 2022

49. Reisinger, Y., Turner, L.: Cultural differences between mandarin-speaking tourists and Australian hosts and their impact on cross-cultural tourist-host interaction. J. Bus. Res. **42**, 175–187 (1998)

50. Tsang, N.K.: Dimensions of Chinese culture values in relation to service provision in hospitality and tourism industry. Int. J. Hosp. Manag. **30**, 670–679 (2011)

51. IFH Köln: Corona Consumer Check Vol. 9. IFH Köln (2021)

52. Bartikowski, B., Laroche, M., Richard, M.-O.: A content analysis of fear appeal advertising in Canada, China, and France. J. Bus. Res. **103**, 232–239 (2019)

53. Jiang, Y., Kim, J., Choi, J., Kang, M.Y.: From clicks to bricks: the impact of product launches in offline stores for digital retailers. J. Bus. Res. **120**, 302–311 (2020)

54. Krippendorff, K.: Content Analysis. An Introduction to its Methodology. Sage, Thousand Oaks, California (2004)

55. Hayes, A.F., Krippendorff, K.: Answering the call for a standard reliability measure for coding data. Commun. Methods Meas. **1**, 77–89 (2007)

56. Brislin, R.W.: Back-translation for cross-cultural research. J. Cross-Cult. Psychol. **1**, 185–216 (1970)
57. Krippendorff, K.: Reliability in content analysis: some common misconceptions and recommendations. Hum. Commun. Res. **30**, 411–433 (2004)
58. Neuendorf, K.A.: The Content Analysis Guidebook, 2nd edn. Sage, London (2016)
59. Holsti, O.R.: Content Analysis for the Social Sciences and Humanities. Addison-Wesley Publ. Co, Reading (1969)
60. Fung Business Intelligence: The Changing Face of China's Retail Market (2014)
61. Li, H.-L., Wang, D.-C.: The influential factors analysis of China's e-commerce development by applying double-layered grey correlation analysis model. In: SSME (2016)
62. Ye, F., Xie, Z., Tong, Y., Li, Y.: Promised delivery time: Implications for retailer's optimal sales channel strategy. Comput. Ind. Eng. **144**, 106474 (2020)
63. Lu, Y., Rucker, M.: Apparel acquisition via single vs. multiple channels: college students' perspectives in the US and China. J. Retail. Consum. Serv. **13**, 35–50 (2006)
64. GIM: ARD/ZDF-Onlinestudie 2021 (2021). https://www.ard-zdf-onlinestudie.de/files/2021/Beisch_Koch.pdf. Accessed 05 May 2022
65. Li, Y., Zhao, H., Yang, Y.: The study on the preferences of customer personal values with Chinese culture background in services. Phys. Procedia **33**, 505–510 (2012)
66. Clark-Gordon, C.V., Bowman, N.D., Goodboy, A.K., Wright, A.: Anonymity and online self-disclosure: a meta-analysis. Commun. Rep. **32**, 98–111 (2019)
67. Kim, J.M., Jun, M., Kim, C.K.: The effects of culture on consumers' consumption and generation of online reviews. J. Interact. Mark. **43**, 134–150 (2018)
68. Twilio: State of Customer Engagement Report 2022. Twilio (2022)

# Will Mobile Payment Change Germans' Love of Cash? A Comparative Analysis of Mobile Payment, Cash and Card Payment in Germany

Sören Aguirre Reid[1]([✉]), Richard Lackes[1], Markus Siepermann[2], and Valerie Wulfhorst[3]

[1] Technische Universität Dortmund, Dortmund, Germany
{soeren.aguirrereid,richard.lackes}@tu-dortmund.de
[2] Technische Hochschule Mittelhessen, Giessen, Germany
markus.siepermann@mni-thm.de
[3] Fachhochschule Südwestfalen, Soest, Germany
wulfhorst.valerie@fh-swf.de

**Abstract.** Despite a global pandemic, it positively increases the intention to use contactless payments. MP usage still falls behind existing alternative payment methods like cash and card. Thus, this paper aims to explore the drivers and inhibitors of MP usage in relation to cash and card payment, which is still an underresearched area in MP. For this, we conducted an online survey (n = 227) in Germany and ran a partial least squares path modelling approach. The results reveal that relative advantage and compatibility drive the intention to use MP compared to cash and card payment. Especially, the result for relative advantage indicates that MP needs some "good" arguments to convince card-preferring users to switch to MP. Interestingly, users regard the disadvantages of MP differently. For instance, perceived switching costs are significantly stronger for MP compared to cash payment but data threat is significantly stronger for MP compared to card payment. The results of the disadvantages of MP confirm Germans' strong "love" for cash payment and that German consumers regard MP as more similar to the card than to cash payment.

**Keywords:** Mobile payment · Switching cost · Diffusion of innovation · Risk · Mobile payment adoption · Cash and card payment

## 1 Introduction

In the early 1990, the first payment systems used cellular mobile communication network elements and devices [1]. The early mobile payment system (MPS) variants required a radio connection between the users' devices and their mobile network operator to pay by call or SMS. However, such technology as radio links for remote payment systems requires continuous network coverage, which is impossible. Therefore, new technologies, like near-field communication (NFC) or QR-codes, came into consideration. Both so-called proximity-based MPS enabled users to use their devices to pay without a radio

Ē. Nazaruka et al. (Eds.): BIR 2022, LNBIP 462, pp. 141–155, 2022.
https://doi.org/10.1007/978-3-031-16947-2_10

connection to a mobile communication network [2]. Since the development of proximity-based MPS solutions, MP has enjoyed growing popularity in many countries like China or the US [3]. Especially the coronavirus accelerated the diffusion of MP. For instance, MP use in the US hit a milestone with a 29% year-over-year growth in 2020 and surpassed 100 million users in 2021 [4]. In stark contrast to the high adoption countries, the adoption of MP in Germany is still in its infancy. By the end of 2019, less than 7 million German adults have used their mobile devices for mobile payment [5]. Even the coronavirus did not change German consumers' preference for cash payment. Only 6% of German consumers started using mobile payment (MP) during the pandemic. During the pandemic, most German consumers switched from cash to contactless card payment [6, 7].

The reason for the lacking adoption of MP is manifold. For instance, German consumers are aware of the advantages of MP (e.g., speed of payment) [8, 9]; they still prefer paying for daily shopping trips in cash, even after decades of card payment options. 15.6 Billion transactions, or 77.9% of all transactions, are paid with cash in the retail sector in Germany [10–12]. Moreover, 64% of German consumers believe paying with MP is too risky [13]. 77% think that MP is not secure [9]. 47% perceive that MP is not secure and are worried that using MP discloses too much information about the user [12]. Additionally, 38% of German consumers think that MP creates so much switching cost (e.g., too much effort to set up or learn) [9, 12].

Therefore, many studies have investigated the acceptance of MP by consumers. However, most former studies only investigated, for instance, what ease of use means at a generic level. But, we do not know what ease of use or usefulness means in relation to existing alternatives from the consumer perspective [14]. Because consumers face different competing payment methods that may prevent consumers from using MP like card payments in Germany [15]. Only four studies addressed this research gap (see next chapter) but limited their comparative analyses to theoretical constructs like the relative advantage of MP [15], the convenience of payment [16], and alternative attractiveness or monetary value [17]. But innovation also needs to be good in relation to the competing options regarding the risk facets and not only regarding their advantages. For instance, physical cash works without registers for users and their transactions in the retail store [18]. This is in stark contrast to the card or MP payment process. Therefore, research should also extend the comparative analysis to the risk facets of MP. Only one study explored the risk of MP compared to cash payment [17]. Hence, our study will contribute to the field of MP and will provide more insight into the comparative analysis of MP compared to cash and card payments regarding the advantages and disadvantages. Summing up, the following research question shall be answered:

*RQ1: How does the perception of drivers of MP differ compared to cash and card payment?*
*RQ2: How does the perception of inhibitors of MP differ compared to cash and card payment?*

In addition, former studies show that the switching cost aspect is also an inhibitor of the diffusion of MP adoption. However, former research conceptualizes the switching cost as a unidimensional construct [17, 19–22], despite that multiple steps exist like evaluation of alternatives or setup-cost [9, 12, 23]. To overcome the purely cost-driven aspect of switching costs (e.g., buying a new smartphone), we conceptualize switching costs as multiple-dimensional constructs to provide new insights.

To answer our research questions, the remainder of this paper is organized as follows: The following section reviews the existing literature in the field of MP adoption and highlights the contribution of this paper in more detail. In the third section, we develop our research model. To test our research model, we surveyed 227 German consumers. The analysis of this survey is presented in Sect. 4. The paper closes with a discussion of the results, the related implications, and the study's limitations.

## 2  Literature Review

The current study contributes to the field of MP adoption in the following ways. First, with regard to the research call for a comparative analysis of MP to existing alternatives [14], only five studies [1, 15–17, 24] followed this call. However, Liébana-Cabanillas et al. [24] compared MP with SMS as a remote payment system while Dahlberg and Öörni [1] compared MP with electronic invoices. Liébana-Cabanillas et al. [24] showed that security, usefulness and subject norms strongly influence MP over SMS payment. Dahlberg and Öörni [1] revealed that compatibility is more important for MP than for electronic invoices. Only Boden et al. [16], Bärsch et al. [15] as well as Loh et al. [17] had a comparative look at more traditional payment methods. Boden et al. [16] analyzed the convenience of the payment and the willingness to pay (WTP) concerning card and MP. They show that MP can increase customers' WTP compared to card payment through greater convenience of payment. In contrast, we investigated further drivers (relative advantages) and inhibitors of MP (data threat, perceived threat). The study of Bärsch et al. [15] compared MP stepwise with cash and card payment concerning the relative advantages. The study showed that self-efficacy and MP threats have a much higher influence when compared to cash than card payment. Our analysis also compared MP stepwise with cash and card payment regarding the relative advantage. In addition, we also compared MP stepwise concerning the data and perceived threats. A recent study by Loh et al. [17] explains the switching intention from cash payment to MP. The study indicated that alternative attractiveness of MP and security and privacy positively affected the switching intention. In contrast, our study extends the comparison of MP with card payment and considers data threat and perceived threat.

Second, we identified six studies that incorporate perceived costs or switching costs in their research model [17, 19–22, 25]. But the majority considers only the financial aspects like costs of use [22], and financial barriers (e.g., headset) [20]. Only two studies added non-financial cost aspects like cognitive costs [17, 25]. However, those studies mainly operationalize costs or switching costs as unidimensional, despite suggestions that multiple dimensions exist like pre-switching search and evaluation costs or post-switching behavioral and cognitive costs [23]. Therefore, our study re-operationalized the construct switching costs with the multidimensional scale by Jones [23]. Furthermore, none of the former studies investigated the relationship between perceived switching costs and perceived risk. But this relationship is important to consider because customers tend to outweigh potential losses (e.g., having to invest time and effort or money) more than potential gains, which leads to an overestimation of the perceived threat and reduces the intention to use MP [26].

## 3  Theoretical Framework and Hypotheses Development

### 3.1  Diffusion of Innovation Theory

For the investigation of the adoption process, two different families of theories are predominant in the IS literature: Studies that are based on Davis' [27] Technology Acceptance Model (TAM) and its successors (e.g., UTAUT) on the one side and those based on Rogers' [28] Diffusion of Innovation (DOI) on the other side. In contrast to the user-centric view of the TAM, the DOI focuses on the adoption process of innovation in a social group. Earlier studies adapted the characteristics of innovations and the market situation and refined a set of constructs (e.g., complexity, compatibility or relative advantages) that could be used to study individual technology acceptance [28, 29]. In particular, many studies in the field of MP support the predictive power of DOI variables like compatibility e.g., [1, 19, 21, 22, 30] or relative advantage for the adoption process [1, 19, 21, 30–32]. Especially, the perceived benefits of a system play an important role in the adoption process. Therefore, this study also considers the relative advantage in the research model. Former studies operationalized the relative advantage with key attributes like the use around the clock [19], usefulness or the speed and efficiency of MP [19, 21, 32].

However, such objective advantages are only one side of the "usefulness coin". People evaluate innovations in comparison to existing payment options; therefore, an innovation must be good in absolute terms and in comparison to existing payment options [15]. The relative advantage can be defined as: *"the degree to which an innovation is perceived as being better than its precursor"* [29, p. 195]. In keeping with the definition, MP has to be compared with the existing payment option. For this, we compare MP with cash and card payment regarding the speed of the payment (e.g., consumers do not need to control the change/entering the pin at the terminal for the card payment) [15, 32], reduced need for carrying a wallet [33]. Further included aspects are the hygienical factor (consumers do not need to touch the money/terminal to enter the pin) [11, 12], the easiness of the payment process (e.g., consumers do not need to count the coins) [21, 30, 34], and the improved overview of account movements [9, 15, 19]. Due to the similarities between MP and card payment, we expect that the "arguments" to convince users to switch to MP need to be stronger [15]. Hence, we hypothesize:

*$H_1$: The relative advantages of MP positively influence the intention to use compared to card payment to a greater extent than cash payment.*

The introduction of MP is not a sure-fire success. MP needs to encompass and reconcile with existing values, behavioural patterns, and experiences of potential users. Extant research confirmed the positive effects of compatibility of innovative technology in general [35] and in particular for MP [1, 19, 21, 22, 30]. The construct compatibility is defined as *"the degree to which an innovation is perceived as being consistent with the existing values, past experiences, and needs of potential adopters"* [28, p. 15]. To address the idea of compatibility, we consider the following aspects: fits well with my lifestyle, the way I manage my finances, the way I pay for products and services [30] and generally fits well with my payment behaviour [36]. Hence, we hypothesize:

*$H_2$: Compatibility of MP positively influence the intention to use MP.*

### 3.2 Theory of Reasoned Action and Theory of Planned Behaviour

The theory of reasoned action (TRA) is one of the most fundamental and influential theories of human behaviour to predict the determinants of consciously intended behaviour [37]. The TRA is very general designed and has been used to predict a wide range of behaviours and the individual acceptance of technology [27, 38]. The goal of the TRA is to understand an individual's voluntary behaviour, like the adoption of MP [39]. The TRA incorporate the attitude toward behaviour and the subjective norm (social influence). Especially social influence was identified as a significant driver for the initial adoption of MP [22, 25, 32, 33]. In particular, studies show that individuals are embedded in a social group and tend to consult their social network to reduce any anxiety regarding the adoption of new technology [33]. Further aspects are what relatives (friends/people around me) think [22, 25, 32] or what the (social) environment expects (e.g., Media, Work, Society) regarding the usage of MP [20]. Therefore, the construct social influence can be defined as: *"the extent to which users perceive that important others (e.g., family and friends) believe they should use a particular technology"* [40, p. 159]. In line with the definition and former studies, our research included the impact of *"people who are important to me"* [32, p. 18] or *"people whose opinions that I value"* [33, p. 214]. Moreover, former findings have shown that the evaluation of new technologies is indirectly shaped by friends, relatives, mass media, society, and the retailer [20]. Hence, we hypothesize:

*$H_3$: Social Influence of MP positively influence the intention to use MP.*

### 3.3 Multidimensional Nature of Switching Cost

To adopt MP, users have to bear the cost because there is no such thing as a free lunch. Therefore, users will conduct a cost-benefit evaluation before deciding [20]. In particular, cash payment is still the dominant way to pay in the retail market. Customers face higher investment costs for MP compared to cash than card payment. For instance, they need to evaluate the existing MP apps or learn how to pay contactless. Former studies partly confirm the cost-benefit evaluation intuition. A few studies reported a significant negative impact on the intention to use [19, 20, 22], whereas other studies did not find any significant relationship between (perceived switching) costs and intention to use [21, 25]. In contrast to the former studies, we consider switching costs as a multiple dimension to address the various evaluation steps of the consumers [23]. Therefore, we incorporate the multidimensional scale of switching costs by Jones et al. [23]. The scale consider: 1. Pre-switching search and evaluation costs: *"perception of the time and effort of gathering and evaluating information prior to switching"* [23, p. 442]; 2. Post-switching behavioral and cognitive: *"perceptions of the time and effort of learning a new service routine subsequent to switching*" [23, p. 442]; and 3. Sunk costs: "*perceptions of investments and costs already incurred in establishing and maintaining relationship*" [23, p. 442]. We excluded the cost of lost performance, uncertainty, and sunk costs because consumers can switch back to cash or card payments anytime.

Furthermore, none of the former studies investigated the relationship between perceived switching costs and perceived risk. However, it should be investigated for two reasons: First, consumers must invest time and effort to set up MP. Some consumers also have to buy a new phone [12]. When consumers find that the costs (investments) outweigh the benefits, they will be unwilling to expend the effort required to switch [17]. Second, consumers tend to weigh potential losses (e.g. investments) as more significant than potential gains when switching to MP. This behaviour is also known as the upward bias, which leads to an overestimation of the perceived threat of a new system [26, 41, 42]. Hence, we hypothesize:

*H₄: The perceived switching cost of MP positively influence the perceived threat of MP compared to cash payment to a greater extent than card payment.*

### 3.4  Data Threat and Perceived Threat

Although previous studies extensively made use of risk items within their research models, the findings are still ambiguous regarding the relation between the constructs of data threat and perceived threat. While Bärsch [15] revealed that data threat has a significant positive impact on the perceived threat, the study of Jenkins & Ophoff [43] did not find a significant impact. Nevertheless, German consumers believe that "giving away too much information" is one of the most severe disadvantages of MP [9]. Therefore, the current study integrated data threat into their research model. Data threat is defined as the "*potential loss of control over personal information, such as when information about you is used without your knowledge or permission. The extreme case is where a consumer is 'spoofed' meaning a criminal uses their identity to perform fraudulent transactions*" [44, p. 455]. We compared MP with cash and card payment regarding the following aspects of data threat: the risk that criminals can access the user account [19, 20, 22, 33], the service provider could send personal information to other companies without the knowledge of the user [21]. Moreover, we incorporate also the aspect of becoming a "transparent customer" (loss of privacy) [15]. Moreover, we expect that the data threat of MP is higher compared to cash than card payment because cash enhances privacy and leaves hardly any traces [18]. Hence, we hypothesize:

*H₅: The data threat of MP positively influence the perceived threat of MP compared to cash payment to a greater extent than card payment.*

MP bears several risks for users, which shape their perception. Therefore it is not surprising that former studies identified perceived risk as an essential factor for adoption [20–22, 25, 32, 33]. The perceived threat can be defined as "the expectation of losses associated with the purchase and acts as an inhibitor to purchase behaviour" [45] p. 185. Thus, this study considers the general risk level of MP compared to cash or card payments [33, 45] and the psychological aspect. Moreover, we also consider performance risk (technical and availability perspective) because the lack of availability of MP compared to existing payment options is among the main inhibitors for German consumers to adopt MP [9, 12]. Moreover, we expect that the perceived threat of MP is higher compared to cash than card payment because cash payment is the legal tender in Germany and bears very low risks. Hence, we hypothesize:

$H_6$: *The perceived threat of MP positively influence the intention to use MP compared to cash payment to a greater extent than card payment.*

The resulting research model is depicted in Fig. 1.

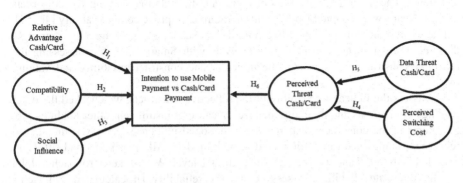

**Fig. 1.** Research model

## 4 Methodology

### 4.1 Data Collection

Our target population comprised consumers in Germany. Each customer answered the questions about the MP usage compared to cash and card payments. This approach aligns with former research [15, 16]. In line with the actual MP usage in Germany, we considered MP as a proximity payment method (NFC and QR code) (Deutsche Bundesbank 2019). The first part of the survey collected data on MP adoption. The second part was focused on demographic variables (e.g., gender and age). We also conducted a survey instrument pre-test to rectify any problems. Using the snowball principle for the survey distribution, we collected 401 responses. Considering the recommendation of Hair et al. [46, 47], 144 observations with more than 15% missing values had to be eliminated, resulting in 257 observations. 227 participants answered the questionnaire completely, beyond the recommended sample size of 40, to receive stable model estimation results [48]. The demographics of the sample show that 51% (130) are female, 44% (113) are male, and 1% (three) are divers. 4% (11) did not provide any gender information. 56% (144) are between 18–34 years old, and 39% (102) are older than > 35 years. 4% (11) did not provide any information regarding their age. We asked our participants to indicate if they had already used MP and which type of MP (QR code or NFC). 64% (165) did not use MP before, and 32% (81) used MP (59 via NFC, 22 via QR-Code). 4% (11) did not provide any information regarding the MP usage experience. In line with former studies in the MP field, e.g. [15, 16] as well as IS research [40], we consider gender and age as control variables in both models. Gender was coded using a dummy variable of 0 (female) or 1 (male). In the case of age, we followed the approach of Liébana-Cabanillas et al. [49] and divided age into two groups ($<35$; $>35$).

### 4.2 Measurement Model

We applied a structural equation modelling approach that consists of an outer and an inner model [46]. The outer measurement model defines the relations between constructs and items. The inner structural model represents the relations among the constructs [50]. All items were adapted from extant literature to improve content validity [46, 51]. All items were measured using a five-point Likert scale ("strongly agree" to "strongly disagree"). We ran the statistical data analysis with SmartPLS 3. For assessing the reflective constructs, we consider the indicator and construct reliability and validity [46].

To assess the indicator reliability of the reflective constructs, we checked the outer loadings of the items and their significance. Because of insufficient outer loadings [46, 52], items in the constructs of perceived cost, social influence, perceived threat, and relative advantage had to be eliminated in both models. All other items had sufficient outer loadings $> 0.7$ and were significant at the 1% level. We checked Cronbach's alpha and the composite reliability to assess the construct reliability. The calculated Cronbach's alpha coefficient exceeds the recommended threshold of 0.7 [53] for all constructs of both models, except for the perceived threat construct in the cash (0.639) and card model (0.699). However, the perceived threat construct exceeds the recommended thresholds of the composite reliability and the AVE in both models. Therefore, we contextualised our measurement of Cronbach's alpha and decided to keep both constructs. Moreover, the composite reliability is more suitable for PLS-SEM, which supports our decision to keep both constructs [46]. The composite reliability of the remaining constructs for both models is higher than 0.7. The AVE of each latent construct in both models exceeds the threshold of 0.5 [54].

Furthermore, all model construct correlations are significant, except for social influence on relative advantage in the card model. For the assessment of the validity, we consider the cross-loadings of the constructs and the Fornell-Larcker criterion. The cross-loadings must exceed all other loadings to the other constructs, which is the case [52]. For the Fornell-Larcker criterion requires, the squared AVE of a construct must be greater than its highest correlation with another construct, which is also the case [50, 52]. Additionally, we used the heterotrait-monotrait ratio of correlations (HTMT) to identify discriminant validity. We selected the $HTMT_{85}$ and $HTMT_{90}$ to assess discriminant value and confirm discriminant validity with an $HTMT_{85}$ and $HTMT_{90}$ of all constructs for both models [55].

### 4.3  Structural Model

To validate the model, we tested for the variance inflation factor (VIFs) of each item and construct to identify potential multicollinearity. We deleted the item COM4 (compatibility) with a value of $5.213 > 5$, the item S1 (social influence) with a value of $5.187 > 5$, and S2 (social influence) with a value of $5.181 > 5$ in both models [46]. The VIF values of constructs ranged from 1.064 to 1.800 in the cash model and from 1.073 to 1.546 in the card model, suggesting that multicollinearity is not a concern [52]. In the next step, we assessed the primary evaluation criteria for the structural model: the $R^2$ level and the significance of the path coefficients. The structural model for MP compared to cash shows a moderate $R^2$ level for MP intention to use with 64.6% and a weak $R^2$ level for perceived threat with 40.5%. Also, for the card model, the $R^2$ level can be considered as moderate with 66.1% for MP intention to use and weak with 44.2% for perceived threat [46].

Furthermore, we quantified how substantial the significant effects are by assessing their effect size $f^2$. The values of their effect size $f^2$ can be described as strong (0.35), moderate (0.15) and weak (0.02) [56]. Additionally, we controlled for a common method bias (CMB) by checking for overlap in items in different constructs in the first step [57]. For instance, in the text for the construct compatibility question, we do not use the word "need" to avoid confounding the construct relative advantages [29]. In the second step, we run Harman's single-factor test with unrotated factor analysis. The result indicates 46.89% of the total variance for a single factor in the cash model (46.45% in the card model), implying that CMB is not substantial in both models (Podsakoff et al. 2003). In the third step, the correlation matrix revealed that all correlations are below 0.74 in both models, while CMB is a problem with high correlations ($r > 0.90$) [54]. Lastly, we also consider the approach of Kock [58]. The results show that all construct VIF value relationships are below 3.3 at the factor level in both models, which indicates that the CMB is not a concern [58]. The bootstrapping analysis of 5000 sub-samples allows for statistical testing of the hypotheses.

### 4.4  Results

To test our hypothesis, we also need to consider a modified t-test (Liébana-Cabanillas et al. 2014; 2017) to check for significant differences in the regression coefficient between both models. We could confirm four out of six hypotheses of our research model (see Table 1). The relation between the relative advantage of MP and the intention to use MP is significant in both models, and the path coefficient for relative advantage is significantly higher in the card model, which confirms our hypothesis $H_1$. The results also support Hypothesis $H_2$ in both models. Moreover, the effect size of Hypothesis $H_2$ is strong in both models. We cannot support Hypothesis $H_3$ in both models. The relation between perceived switching cost and the perceived threat is significant in both models, and the path coefficient is significantly stronger in the cash model, which confirms our hypothesis $H_4$. Both models show a significant relationship between data threat and perceived threat. However, the relationship is significantly stronger for the card model. Therefore, we can only partially confirm our hypothesis $H_5$. The relationship between perceived threat and

**Table 1.** Path coefficients, $f^2$ values and significant differences for the cash and card model.

| Hypothesis | Path coefficient | | | | Differences path coefficients |
| --- | --- | --- | --- | --- | --- |
| | Cash Model | $f^2$ Cash Model | Card Model | $f^2$ Card Model | |
| $H_1$ confirm | 0.232** | 0.098 | 0.323** | 0.237 | $-0.091$** |
| $H_2$ confirm | 0.536** | 0.451 | 0.516** | 0.507 | $0.021^{ns}$ |
| $H_3$ not confirm | $0.082^{ns}$ | 0.017 | $0.027^{ns}$ | 0.002 | - |
| $H_4$ confirm | 0.207** | 0.065 | 0.128* | 0.025 | 0.079** |
| $H_5$ partial confirm | 0.542** | 0.446 | 0.605** | 0.561 | $-0.064$** |
| $H_6$ confirm | $-0.212$** | 0.092 | $-0.177$** | 0.070 | $-0.034$* |

ns: non-significant, 1% level**, 5% level*; $f^2$: strong (0.35), moderate (0.15), weak (0.02)

intention to use is significant in both models, and the relationship is significantly stronger for the cash model, which confirms our hypothesis $H_6$.

Regarding control variables, only age significantly influences the intention to use in both models (0.103** cash model; 0.098** card model).

## 5  Discussion

The present study examined two central research questions: (*RQ1*) How does the perception of drivers of MP differ compared to cash and card payment, and (*RQ2*) how does the perception of inhibitors of MP differ compared to cash and card payment?

With regard to *RQ1*, our study confirms the decisive role of the relative advantage of MP for convincing cash and card preferring users in general [16, 32, 33]. The results show that the relative advantage of MP is more vital compared to the card than cash payments. This result makes sense because MP and card payment processes have similarities, like the fast and easy payment process (e.g., no money counting) [15]. Therefore, MP needs some "good" arguments to convince users to switch from card payment to MP. The compatibility of users with MP is the strongest driver for the MP intention to use. The impact and strength of the result of compatibility are in line with previous findings [21, 22, 30]. Therefore, it is essential for users that MP fits users' way they manage their finance or pay for products and services. Interestingly, the structural model revealed that social influence is not essential for users to start using MP in both models. These findings contrast the existing research [22, 25, 32]. This finding indicated that users are not influenced by their environment, society or the retailer. An explanation could be the low usage rate by other consumers.

With regard to the *RQ2*, none of the former studies investigated the relationship between perceived switching costs and perceived threat [19, 20, 22]. The findings revealed a significant impact of the perceived switching costs on the perceived threats of MP compared to cash and card payments. Moreover, the finding for MP compared to cash is significantly stronger, which confirms Germans' strong "love" for cash payment as the incumbent payment option [10, 12, 59]. Therefore, German customers are more unwilling to expend their effort to switch and form an increased perceived threat toward using MP [17]. Whereas the perceived switching cost for MP compared to card payment is significantly lower. An explanation could be the similarity of the payment process of both payment options (e.g., contactless). In addition, the outer loadings of the items pre-switching search and evaluation costs and cognitive costs of the construct switching cost are significantly relevant in both models, except for the item setup costs, which are not relevant. The results also show that data threat significantly increases the perceived risk of MP compared to cash and card payments. The finding is in line with Bärsch et al. [15] but in contrast with Jenkins and Ophoff [43]. Moreover, the result revealed data threat is significantly stronger for MP compared to card payment. It seems that users are more afraid of "giving away too much information" when MP is compared to the card. An explanation for the higher data risk could be that users have to enter a lot of information into their smartphone or MP app (e.g., card numbers, bank account details and other personal information) to use MP. In addition, the finding extended the former research results and revealed that customers' data threats differ regarding comparing MP with cash or card payments. Therefore it is important to analyze technology adoption not only one absolute level but also on the relative level (e.g., comparison with existing alternatives). The estimated results also confirmed the significant negative impact of perceived threats on the intention to use MP, which is in line with previous research [20, 21, 25, 33]. The results also show that the perceived threat of MP is significantly stronger compared to cash than card payment. The result is in line with our expectations and confirms that people are used to paying with cash as the most common payment option in Germany. Therefore, people regard MP as more similar to the card than cash payment and are still stuck to cash payment, so they regard MP as much more critical and risky [15].

### 5.1 Managerial Implications

First, practitioners need to promote MP compared to cash and card payment differently. Users who compared MP with cash payment are less susceptible to objective advantages and focus more on the compatibility of MP. Users who compared MP with card payment are more susceptible to objective advantages than compatibility of MP.

Second, practitioners need to advertise the compatibility of MP in general. Practitioners should tailor their marketing measures accordingly so that MP is perceived as compatible with German users' lifestyles. Thirdly, practitioners should also use their marketing effort to explain how easy it is to adopt MP to reduce the perceived switching cost compared to cash and card payments. In particular, it is essential for cash preferring users to make the switch from cash to MP as simple as possible. For instance, provide an operating manual in simple language to reduce the cognitive effort for MP's installation and learning process. Fourthly, privacy plays a vital role in adopting decisions for German users. In particular, users are not likely to trust companies like Apple or Payback Pay compared to their bank institutes [8]. Therefore, practitioners need to highlight the security of MP in general. For instance, practitioners should signal and explain security measures to protect bank accounts or users' identities. Moreover, they should transparently communicate the collected data and the data processing.

### 5.2  Limitation and Future Research

First, the collected data comprises only German-speaking users of younger age. However, young people have been proven to be adequate surrogates for decision-makers [60], so that this sampling hardly distorts the results. Future research should collect a more representative and also enlarge the sample size in different cultural regions. Second, our work only examines the drivers and inhibitors of MP compared to cash and card payments based on questionnaire statements and not the actual decision. Future research may consider replicating our findings with another research design to provide further insights with an experiment, interviews or actual data. Third, we do not split the sample based on users' prior experience (self-report) and analyze the difference between inexperienced and experienced users. Hence, future research should consider users' experience level, like how long a user has already used MP and its effect on the perception of driver and inhibitors of MP. Fourth, although the results indicate that users value data privacy, the reality shows that they also ignore privacy in their everyday use (e.g., social media). Therefore, it remains unclear what exactly does full transparency mean? Do digital consumers want full transparency? And how does full transparency affect the interaction with the digital services (more data processing dialogue boxes). Future research needs to address these questions in the context of MP and other human-computer interactions.

## References

1. Dahlberg, T., Oorni, A.: Understanding changes in consumer payment habits - do mobile payments and electronic invoices attract consumers? In: 40th HICCS, Hawaii, pp. 1–10. IEEE (2007)
2. Gerpott, T.J., Meinert, P.: Who signs up for NFC mobile payment services? Mobile network operator subscribers in Germany. Electron. Commer. Res. Appl. **23**, 1–13 (2017)
3. eMarketer. https://www.emarketer.com/content/china-mobile-payment-users-2019. Accessed 17 May 2022
4. eMarketer. https://www.emarketer.com/content/us-payment-users-will-surpass-100-million-this-year. Accessed 27 July 2022
5. eMarketer. https://www.emarketer.com/content/germany-mobile-payment-users-2019. Accessed 27 July 2022
6. IFH Köln. https://www.ifhkoeln.de/shop-7kategorie/e-commerce/. Accessed 29 Apr 2022
7. IFH Köln. https://www.ifhkoeln.de/paymenttrend-nutzung-des-smartphones-zur-bezahlung-immer-selbstverstaendlicher/. Accessed 29 Apr 2022
8. Bitkom. https://www.bitkom-research.de/de/system/files?file=document/Bitkom-Pr%C3%A4sentation%20PK%20SID%202019.pdf. Accessed 21 May 2022
9. PWC. https://www.pwc.de/de/digitale-transformation/pwc-studie-mobile-payment-2019.pdf. Accessed 23 May 2022
10. Cabinakova, J., Knümann, F., Horst, F.: Kosten der Bargeldzahlung im Einzelhandel. Studie zur Ermittlung und Bewertung der Kosten, die durch die Bargeldzahlung im Einzelhandel verursacht werden, Deutsche Bundesbank (2019)

11. Deutsche Bundesbank: Zwischenerhebung zum Zahlungsverhalten in Deutschland 2019, Deutsche Bundesbank (2019)
12. Pietrowiak, A., Korella, L., Novotny, J.: Zahlungsverhalten in Deutschland 2020 – Bezahlen im Jahr der Corona-Pandemie. Erhebung über die Verwendung von Zahlungsmitteln, Deutsche Bundesbank (2021)
13. EHI Retail. https://www.mobile-payment-initiative.de/mobile-payment/verbraucherbefr agung-2018/. Accessed 27 Apr 2022
14. Dahlberg, T., Guo, J., Ondrus, J.: A critical review of mobile payment research. Electron. Commer. Res. Appl. **14**, 265–284 (2015)
15. Bärsch, S., Siepermann, M., Lackes, R., Wulfhorst, V.: Nothing but cash? Mobile payment acceptance in Germany. In: 41th ICIS, pp. 1–17. AIS, India (2020)
16. Boden, J., Maier, E., Wilken, R.: The effect of credit card versus mobile payment on convenience and consumers' willingness to pay. J. Retail. Consum. Serv. **52**, 1–10 (2020)
17. Loh, X.-M., Lee, V.-H., Tan, G.W.-H., Ooi, K.-B., Dwivedi, Y.K.: Switching from cash to mobile payment: what's the hold-up? INTR **31**, 376–399 (2021)
18. Mai, H.: Cash empowers the individual through data protection, DB Research (2019)
19. Lu, Y., Yang, S., Chau, P.Y., Cao, Y.: Dynamics between the trust transfer process and intention to use mobile payment services: a cross-environment perspective. Inf. Manag. **48**, 393–403 (2011)
20. Hongxia, P., Xianhao, X., Weidan, L.: Drivers and barriers in the acceptance of mobile payment in China. In: ICEE, Shanghai, pp. 1–4. IEEE (2011)
21. Pham, T.-T.T., Ho, J.C.: The effects of product-related, personal-related factors and attractiveness of alternatives on consumer adoption of NFC-based mobile payments. Technol. Soc. **43**, 159–172 (2015)
22. Yang, S., Lu, Y., Gupta, S., Cao, Y., Zhang, R.: Mobile payment services adoption across time: an empirical study of the effects of behavioral beliefs, social influences, and personal traits. Comput. Hum. Behav. **28**, 129–142 (2012)
23. Jones, M.A., Mothersbaugh, D.L., Beatty, S.E.: Why customers stay: measuring the underlying dimensions of services switching costs and managing their differential strategic outcomes. J. Bus. Res. **55**, 441–450 (2002)
24. Liébana-Cabanillas, F., Ramos de Luna, I., Montoro-Ríos, F.: Intention to use new mobile payment systems: a comparative analysis of SMS and NFC payments. Econ. Res.h-Ekonomska Istraživanja **30**, 892–910 (2017)
25. Abrahão, R.D.S., Moriguchi, S.N., Andrade, D.F.: Intention of adoption of mobile payment: an analysis in the light of the unified theory of acceptance and use of technology (UTAUT). RAI Revista de Administração e Inovação **13**, 221–230 (2016)
26. Kim, H.-W., Kankanhalli, A.: Investigating user resistance to information systems implementation: a status quo bias perspective. MIS Q. **33**, 567–582 (2009)
27. Davis, F.D.: Perceived usefulness, perceived ease of use, and user acceptance of information technology. MIS Q. **13**, 319–340 (1989)
28. Rogers, E.M.: Diffusion of Innovations, 4th edn. The Free Press, New York (2010)
29. Moore, G.C., Benbasat, I.: Development of an instrument to measure the perceptions of adopting an information technology innovation. Inf. Syst. Res. **2**, 192–222 (1991)
30. Schierz, P.G., Schilke, O., Wirtz, B.W.: Understanding consumer acceptance of mobile payment services: an empirical analysis. Electron. Commer. Res. Appl. **9**, 209–216 (2010)
31. Bailey, A.A., Pentina, I., Mishra, A.S., Ben Mimoun, M.S.: Mobile payments adoption by US consumers: an extended TAM. IJRDM **45**, 626–640 (2017)
32. Koenig-Lewis, N., Marquet, M., Palmer, A., Zhao, A.L.: Enjoyment and social influence: predicting mobile payment adoption. Serv. Ind. J. **35**, 537–554 (2015)
33. Slade, E., Williams, M., Dwivedi, Y., Piercy, N.: Exploring consumer adoption of proximity mobile payments. J. Strateg. Mark. **23**, 209–223 (2015)

34. Ramos de Luna, I., Liébana-Cabanillas, F., Sánchez-Fernández, J., Muñoz-Leiva, F.: Mobile payment is not all the same: the adoption of mobile payment systems depending on the technology applied. Technol. Forecast. Soc. Change **146**, 931–944 (2019)
35. Tornatzky, L.G., Klein, K.J.: Innovation characteristics and innovation adoption-implementation: a meta-analysis of findings. IEEE Trans. Eng. Manag. **29**, 28–45 (1982)
36. Oliveira, T., Thomas, M., Baptista, G., Campos, F.: Mobile payment: understanding the determinants of customer adoption and intention to recommend the technology. Comput. Hum. Behav. **61**, 404–414 (2016)
37. Ajzen, I., Fishbein, M.: Understanding Attitudes and Predicting Social Behavior. Pearson, New Jersey (1980)
38. Sheppard, B.H., Hartwick, J., Warshaw, P.R.: The theory of reasoned action: a meta-analysis of past research with recommendations for modifications and future research. J. Consum. Res. **15**, 325–343 (1988)
39. Ajzen, I.: The theory of planned behavior. Organ. Behav. Hum. Decis. Process. **50**, 179–211 (1991)
40. Venkatesh, V., Thong, J., Xu, X.: Consumer acceptance and use of information technology: extending the unified theory of acceptance and use of technology. MIS Q. **36**, 157–178 (2012)
41. Lin, T.-C., Huang, S.-L., Hsu, C.-J.: A dual-factor model of loyalty to IT product – the case of smartphones. Int. J. Inf. Manag. **35**, 215–228 (2015)
42. Polites, G.L., Karahanna, E.: Shackled to the status quo: the inhibiting effects of incumbent system habit, switching costs, and inertia on new system acceptance. MIS Q. **36**, 21 (2012)
43. Jenkins, P., Ophoff, J.: Factors influencing the intention to adopt NFC mobile payments – a South African perspective. In: CONF-IRM, pp. 1–12. AIS (2016)
44. Featherman, M.S., Pavlou, P.A.: Predicting e-services adoption: a perceived risk facets perspective. Int. J. Hum. Comput. Stud. **59**, 451–474 (2003)
45. Peter, J.P., Ryan, M.J.: An investigation of perceived risk at the brand level. J. Mark. Res. **13**, 184–188 (1976)
46. Hair, J.F., Ringle, C.M., Sarstedt, M.: PLS-SEM: indeed a silver bullet. J. Mark. Theory Pract. **19**, 139–152 (2011)
47. Hair, J.F., Hult, G.T.M., Ringle, C.M., Sarstedt, M., Richter, N.F., Hauff, S.: Partial Least Squares Strukturgleichungsmodellierung. Eine anwendungsorientierte Einführung. Verlag Franz Vahlen, München (2017)
48. Chin, W.W.: The partial least squares approach to structural equation modeling. Mod. Methods Bus. Res. **295**, 295–336 (1998)
49. Liébana-Cabanillas, F., Sánchez-Fernández, J., Muñoz-Leiva, F.: Antecedents of the adoption of the new mobile payment systems: the moderating effect of age. Comput. Hum. Behav. **35**, 464–478 (2014)
50. Fornell, C., Larcker, D.F.: Structural equation models with unobservable variables and measurement error. J. Mark. Res. **18**, 39–50 (1981)
51. Straub, D., Gefen, D., Boudreau, M.-C.: Validation guidelines for IS positivist research. CAIS **13**, 380–427 (2004)
52. Hair, J.F., Sarstedt, M., Hopkins, L.G., Kuppelwieser, V.: Partial least squares structural equation modeling (PLS-SEM) - an emerging tool in business research. Eur. Bus. Rev. **26**, 106–121 (2014)
53. Nunnally, J.C.: Psychometric Theory. McGraw-Hill, New York (1978)
54. Bagozzi, R.P., Yi, Y.: On the evaluation of structural equation models. JAMS **16**, 74–94 (1988)
55. Henseler, J., Ringle, C.M., Sarstedt, M.: A new criterion for assessing discriminant validity in variance-based structural equation modeling. J. Acad. Mark. Sci. **43**(1), 115–135 (2014). https://doi.org/10.1007/s11747-014-0403-8
56. Cohen, J.: Statistical Power Analysis for the Behavioral Sciences, 2nd edn. Routledge, New York (1988)

57. Conway, J.M., Lance, C.E.: What reviewers should expect from authors regarding common method bias in organizational research. J. Bus. Psychol. **25**, 325–334 (2010)

58. Kock, N.: Common method bias in PLS-SEM. Int. J. e-Collab. **11**, 1–10 (2015)

59. Onlinemarktplatz. https://www.onlinemarktplatz.de/111278/ehi-studie-liebe-zum-bargeld-laesst-nach/?amp. Accessed 24 July 2022

60. Remus, W.: Graduate students as surrogates for managers in experiments on business decision making. J. Bus. Res. **14**, 19–25 (1986)

# Digital Business

# Drawing Attention on (Visually) Competitive Online Shopping Platforms – An Eye-Tracking Study Analysing the Effects of Visual Cues on the Amazon Marketplace

Alper Beşer[✉] , Julian Sengewald , and Richard Lackes

Technische Universität Dortmund, 44227 Dortmund, Germany
alper.beser@tu-dortmund.de

**Abstract.** Amazon has become the market leader among online shopping platforms. Potential customers can search for products on Amazon and compare different offers. However, the highly (visually) competitive marketplace can make it difficult for sellers to stand out from the crowd. In an eye-tracking experiment, we investigate how visual cues (e.g. "Bestseller" badge) influence customers' behaviour by attracting attention, and how they affect their product choice. The experiment with a sample size of $N = 60$ participants was conducted on a German university campus. With the obtained eye-tracking data, we use a lognormal mixed-effects model and perform a logistic regression for estimating the effects of visual cues on Amazon search pages. The results indicate that visual cues marginally influence the viewing duration and decision time of customers but can have a considerable impact on the product choice.

**Keywords:** E-commerce · Online shopping platforms · Customer engagement · Customer attention · Eye-tracking

## 1 Introduction

E-commerce and online shopping have become increasingly popular [1]. Consumers prefer e-commerce to in-store purchases because it is more affordable and convenient (e.g. due to location independence and easier comparisons of different product offers and sellers) [2, 3]. Although e-commerce provides sellers with a broad toolset, it also possesses unique challenges. For instance, distraction in the digital world is a greater risk for e-commerce marketing than for traditional marketing in physical stores. Not only are offers of competing sellers easily accessible via a mouse click, but there is also the risk of diverting customers' attention by poorly designed websites (e.g. confusing navigation or misleading presentation of information). Therefore, customers' attention has become a valuable commodity in the online context, and online retailers continuously seek new strategies to stand out from the crowd. One way for drawing attention is visual salience [4]. Different strategies for visual attention seeking by using visual cues have evolved in online marketing [5, 6]. However, the risk of overdoing is a known concern. In print

© The Author(s), under exclusive license to Springer Nature Switzerland AG 2022
Ē. Nazaruka et al. (Eds.): BIR 2022, LNBIP 462, pp. 159–174, 2022.
https://doi.org/10.1007/978-3-031-16947-2_11

advertising, for instance, exaggerated advertisement design is called competitive clutter [7]. Banner blindness is a phenomenon in the online context where in situations with excessive visual stimuli the perceptual filtering system of customers protects them from information overload, which basically makes customers overlook the advertisement and ignore it [8].

Amazon, as being Europe's largest online shopping platform [9, 10], is also affected by such competitive clutter. Amazon's popularity has turned it into a highly competitive marketplace that attracts a multitude of sellers [11, 12]. When customers search for products, they often find many equivalent options for purchase that are also visually very similar (see Fig. 1 for an example). Such (visually) competitive environments lead to an information overload that negatively affects customers' purchase decision-making process [13–15].

**Fig. 1.** Example of an Amazon search page, which can be broadly categorised into four areas. The search bar (box 1) contains the search query, for which the results are presented in the main content area of the page (box 2). The sponsored brand area (box 3) can be used by brands to advertise their logo with a custom headline and multiple products [16]. The left side of the search page (box 4) provides the user with advanced filtering functions for the search query.

To address this, Amazon provides decision-making aids by adding visual cues to the most popular products in terms of recent sales ("Bestseller") or noteworthy choices ("Amazon's Choice") [17]. When customers search for products, these badges are not always attached to the products with the highest rank in the search results but can also be placed on lower-ranked offerings. The equipment of product offerings with such visual cues is not unbiased but influenced by various factors (see Sect. 4.2). While these are for the most part controlled by Amazon and can hardly be influenced by marketplace participants [17, 18], there are some ways sellers can alter the presentation of their products in order to cut through the clutter. For instance, sellers could make the product presentation in the offer photo more appealing or offer a discount voucher. However, since the product offers presented on the Amazon marketplace are highly competitive in terms of visual salience, it is not clear if and to which extent the placement of single visual cues can increase the attention for a single offer and ultimately lead to its selection. To shed light on this, we aim to investigate the following research question (RQ) in this study:

**RQ:**    How do visual cues influence customer behaviour on the Amazon marketplace?

To answer the research question, we carried out an eye-tracking experiment and presented test subjects with original and modified Amazon product search pages with the task of selecting the most appealing/interesting product that they most likely would buy. Based on the eye-tracking data, we use a mixed-effects model and logistic regression for estimating the effects of different kinds of visual cues (seller-controlled vs. not seller-controlled) on the customer behaviour represented by the viewing duration of offers, product choice, and the corresponding decision time.

The remainder of this paper is organised as follows. In the next section, we give an overview of related literature and point out our contribution. In the third section, hypotheses regarding customer behaviour on the Amazon marketplace are developed. In the fourth section, we present the applied research method and give details on the (un-)modified Amazon search pages the study participants were shown. The hypotheses are addressed and discussed in the fifth section. The paper closes with a conclusion that includes managerial implications, limitations, and future research directions.

## 2    Related Literature

Eye-tracking can visualise cognitive attention processes [19, 20]. Eye-tracking has gained a foothold in several areas of research such as neuroscience, (cognitive) psychology, marketing/advertising, and human-computer interaction [21–24]. As a sub-stream of the latter research area, eye-tracking is commonly used for user experience research, e.g. on websites [4, 13, 25–27], which our paper is closely related. The proper placement of website items can improve the user's surfing experience [28, 29]. The design and layout of websites are also important in the context of e-commerce and online shopping, where eye-tracking can be used for product attraction measurement [4, 30, 31]. Thereby, eye-tracking data can help to understand how visual attention draws the success of products in e-commerce [23]. Thus, for maintaining profitability and competitiveness, eye-tracking can be used to adapt product offers to market requirements on e-commerce websites [31].

The literature closely related to our study has shifted from the study of the eternity of website design (i.e. the page layout including the use of links and animation) [4], to the study of elements at a lower level of hierarchy. These include product listening pages [4, 32], the design of product pictures [33], and the use of advertisement banners [5, 6, 34]. In addition, a variety of theories have been leveraged for understanding how visual cues affect user behaviour in the online context including the cognitive load theory for visual attention [4, 34], stimulus-organism-response (S-O-R) theory [35], elaboration

likelihood model [36], and theories based on neuro-physical and -psychological literature [5, 6]. We contribute to the user experience research on e-commerce websites by
increasing the level of granularity of the stimulus material. Instead of varying the page
layout or banners, we modify elements on a product-offer level on Amazon search pages
and investigate how customers' attention can be directed towards certain products in
a highly visually competitive environment where competitors try to outdo each other.
Diverging from the above-mentioned studies, which mostly use self-designed website
mock-ups for their experiments, we rely on actual screenshots of Amazon search pages
for testing the effects of modified visual cues on shopping behaviour under realistic
conditions.

## 3   Theoretical Background and Hypothesis Development

The S-O-R model posits that a perceptible stimulus (S) leads to the cognitive decoding
of the presented stimulus by the recipient's organism (O). After the cognitive processing
of the stimulus has been done, there occurs a measurable response (R) of the organism.
Such a response may, for instance, magnify itself in increased product/brand awareness
or purchase intention. For a meta-review of applications of the S-O-R model, see [37]. In
our experiment, we manipulate stimuli (S) on Amazon search pages in order to measure
the response (R) of online shoppers by eye-tracking for deriving conclusions about the
inner mechanisms taking place in the organism (O), i.e. their cognitive behaviour for
which eye movements are indicative [38, 39]. These may then be utilised to deduce
managerial implications regarding the applicability of the different tested stimuli for
improving online shoppers' behavioural responses.

Humans allocate their gaze for purposely processing or searching for information or
because something is visually interesting [40–42]. Consequently, the viewing duration
of objects of interest is increased [4, 43–45]. This mechanism has led to much research
considering interventions for manipulating visual attention (e.g. 4, 46–48). Based on
this, we formulate the first hypothesis:

> **Hypothesis 1:**   Visual cues increase the total viewing duration of the
> corresponding product.

Krajbich et al. [49] postulated the attentional drift diffusion model that links attention
and choice by a mathematical relationship. The underpinning idea is that attention drives
choice, i.e. the aspects that draw more attention are higher valued in the perceived value of
a product [46]. We thus hypothesise that the attention drawn by visual cues facilitates the
decision-making process of customers and makes it more likely that customers choose
a particular product:

**Hypothesis 2:**   Visual attention due to visual cues leads to a higher probability of choosing the corresponding product.

Overall, the view that humans do not process all available information is shared among economists from different fields (e.g. marketing or behavioural economics). For example, the idea that consumers rarely search on the whole set of possible choices has been discussed in the marketing literature as a consideration set [50–52]. In this context, the theory of rational inattention [53] assumes that building the perceived value of a product causes costs due to the searching and processing of information. Given multiple product choices, a customer needs to invest search costs for the evaluation of each product. We hypothesise that visual cues such as the "Bestseller" badge on Amazon search pages constitute an exogenous reduction of search costs. Because the search costs are reduced, the overall decision-making process of customers across all product alternatives on a search page can be accelerated in time:

**Hypothesis 3:**   Visual cues reduce the decision time of customers.

## 4   Research Method

### 4.1   Experimental Setup

The eye-tracking study was carried out on a German university campus. During class breaks, randomly picked students were invited to participate in the experiment. Participants were asked to enter an empty room with a dedicated computer and a 24-inch monitor where the eye-tracking equipment (Gazepoint GP3) was installed. After assisted calibration of the eye-tracking hardware, participants were instructed that they would be shown different Amazon search pages during the experiment. For each participant, the pages were shown in randomised order. To avoid an initial adaptation/familiarisation period that could have distorted the eye-tracking results on the first shown search page (i.e. participants inspect the first page more closely than the following ones), a printout of an exemplary search page was shown to the participants before the experiment was started. The exemplary search page was not included in the set of pages presented during the experiment. The participants were asked to emulate a realistic usage behaviour, i.e. inspect the shown screenshots of the Amazon search pages as if they were browsing them at home, and were also informed that there was no time restriction. The task consisted of choosing the most interesting or appealing product by clicking on it with the mouse. Afterwards, by pressing the space key on the keyboard, the next screenshot of an Amazon search page was presented to the viewer. On each page, the eye movements, viewing durations, and mouse clicks of the participants were recorded. After completing the experiment, the participants were asked to fill out an optional questionnaire with demographic data. All test subjects provided information regarding their age and gender.

## 4.2   Material and Modified Stimuli

When customers search on the Amazon marketplace for a certain product by entering keywords, the results are presented on the Amazon search page, which is structured in several website sections and elements. Figure 1 depicts an exemplary search query for a product used in the eye-tracking experiment. In our experiment, we modified several parameters of products that are listed in the main content area, i.e. box 2 of the search page. The information presented on a product level, as shown in Fig. 2, can be differentiated into two categories: (1) parameters directly controlled by the seller and (2) parameters that are not or only indirectly controlled by the seller. The first category comprises the product photo, title, price, discount vouchers, and the Amazon Prime logo, which is available if the seller participates in the Fulfilment by Amazon (FBA) program for simplified logistic processes that are handled by Amazon. Amazon has strict rules for the modification and placement of text in photos but allows the adding of test winner information [54].

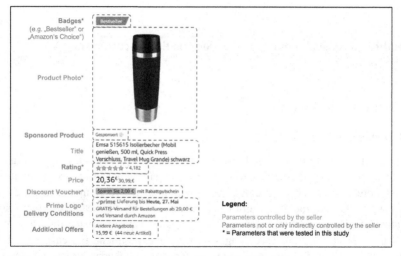

**Fig. 2.**   Categorisation of parameters of a product offer presented on an Amazon search page.

The second category includes the badges (e.g. "Bestseller" or "Amazon's Choice") above the product photo, sponsored production declaration below the photo, product ratings, delivery conditions, and additional offers. The "Bestseller" badge is attached to products that generate the most sales and are refreshed on an hourly basis [18]. The availability of the "Amazon's Choice" badge depends on an undisclosed algorithm that incorporates ratings, price, and shipping availability [17]. Sellers can increase the ranking of their products on search pages by promoting their offers as sponsored products. These are cost-per-click (CPC) ads where sellers only pay when users click on their product offer. The higher the seller's bid on a CPC ad for a product search keyword is, the more likely it is that the offered product obtains higher ranks in the listing [55]. The delivery conditions, such as shipping costs and delivery date, are determined by Amazon if the seller is part of the FBA program [56]. If the seller does not participate in FBA, these become seller-controlled parameters. The additional offers can only be influenced by the seller if s/he owns the brand rights on the offered products. These would entitle the seller to be the only distributor of those particular products on the Amazon marketplace and thereby inhibit other offers.

The set of Amazon search pages consisted of actual search queries on the Amazon marketplace for everyday products that the target subject group (i.e. students on the campus) would most likely be interested in irrespective of their gender (see Table 1). The search pages were manually adjusted with a photo manipulation program. The adjusted parameters constitute the visual cues whose effects on customer behaviour were to be tested. The goal of the experimental setup was to mimic a realistic shopping environment the participants were likely used to. To avoid biased results due to learning and memorisation processes, the original and modified search pages were not shown to the same subjects. Instead of a within-subjects approach, we therefore chose a between-subjects design with two groups for our study. Table 1 lists the product search pages with the modified parameters and the group allocation of their (un-)modified versions. The first three products constitute the set of search pages where the adjusted parameters are directly controlled by the seller. Accordingly, the last three products represent the set of pages where parameters were adjusted that are not or only indirectly controlled by the seller.

**Table 1.** Tested Amazon search pages with parameter adjustments.

| Amazon product search page (Search keywords translated from German) | Rank of the modified product on the search page | Parameter adjustment for adding or removing visual cues | Test subject group A | Test subject group B |
|---|---|---|---|---|
| Thermo mug | 1/6 | Removed the "Prime" logo from the delivery conditions | Unmodified page | Modified page |
| Cooling towel | 6/6 | Added discount voucher | Modified page | Unmodified page |
| Electric toothbrush | 4/6 | Added a certified "Test Winner" logo to the product photo | Modified page | Unmodified page |
| Water bottle | 5/6 | Reduced the product rating from 4.5 to 2 stars | Modified page | Unmodified page |
| Gym bag | 6/6 | Added "Bestseller" badge | Unmodified page | Modified page |
| Hand luggage | 6/6 | Removed "Amazon Choice" badge | Unmodified page | Modified page |

# 5  Results

## 5.1  Dataset Description

The eye-tracking study was carried out during a period of 16 days. In total, $N = 60$ test subjects aged between 18 and 42 years, with an average of 23.2 years, took part in the experiment. 14 persons were female and 46 were male. They were split equally among the two test subject groups, i.e. each group consisted of 7 female and 23 male test subjects. All participants were either students or (uninvolved) employees of the university. A prerequisite for participating in the experiment was a sufficient level of visual acuity which potential test subjects were asked about. 93.3% stated that they had already purchased products on the Amazon marketplace and thus were familiar with the website. 56.7% had an active Amazon Prime membership and therefore most likely knew of the service benefits (e.g. faster shipping) they were eligible for.

## 5.2  Analysis

For carrying out quantitative statistical analyses, we defined areas of interest (AOI) for each of the modified products in Table 1 in the eye-tracking software. These were used for collecting user behavioural data that only concerned the rectangular region on the search page specified by the AOI [26, 57], which then served as the basis for the conducted tests [58].

Given that each of the 60 test subjects was shown 6 different Amazon search pages with a measurement of 6 product positions on each page, we would have a nominal sample size of 36 observations per subject and 2.160 observations in total. However, we had to remove the 36 observations of one test subject as s/he did not seem to cooperate with the experimental condition because the overall response time was less than 2 s on average. Selected observations of other subjects were removed as well if the fixation of an AOI was not sufficiently long enough, leading to the removal of another 43 observations. In total, 79 observations from 27 different participants were removed, reducing the sample size to 2.081.

Note that the observation sample size is smaller for the decision time because it accumulates the viewing durations of all 6 product positions on a page. Thus, the nominal sample size for the decision time is 360, from which 1 was removed due to the above-mentioned reasons. Based on the cleansed sample, we conducted a simple randomisation check regressing the viewing duration on the test condition of the subject groups A and B. No coefficient was significant, confirming that the randomisation worked indeed as intended.

As the viewing duration of an AOI is a strictly positive variable, we model the dependent variable using a generalised linear model (GLM) with a lognormal link function for testing Hypothesis 1. Furthermore, we use a hierarchical lognormal GLM because hierarchical models deal better in situations with repeated measurements. Subjects and products are treated as random effects, while the existence of the visual cue and the demographical data of the test subjects represent the fixed effects of our specification. This procedure is also used for testing Hypothesis 3. In order to verify Hypothesis 2 regarding the product choice of customers, we use a logistic regression. Mixed-effects modelling seems not appropriate for Hypothesis 2 because of our experimental setup, where users always had to make a choice on a presented search page. We also tested a mixed effects specification for logistic regression, but the corresponding random effects resulted in being negligible, and convergence issues in the software fitting procedure arose. Therefore, we prefer both for statistical and subject matter reasons the specification used in Eq. (2).

$$Viewing\ Duration_{subject,page,position}$$
$$= \beta_0 + \beta_1\ Visual\ Cue_{subject,page,position} + \beta_2\ Age \tag{1}$$
$$+ \beta_3\ Gender + \varepsilon_{subject} + \varepsilon_{page,position} + \varepsilon_{subject,page,position}$$

$$Product\ Choice_{subject,page,position}$$
$$= \beta_0 + \beta_1\ Visual\ Cue_{subject,page,position} \tag{2}$$
$$+ \beta_2\ Age + \beta_3\ Gender + \varepsilon_{subject,page,position}$$

$$Decision\ Time_{subject,page}$$
$$= \beta_0 + \beta_1\ Visual\ Cue_{subject,page} \tag{3}$$
$$+ \beta_2\ Age + \beta_3\ Gender + \varepsilon_{subject} + \varepsilon_{subject,page}$$

With:

$$\varepsilon_{subject} \sim N\left(0, \sigma_{subject}\right)$$

$$\varepsilon_{page,position} \sim N\left(0, \sigma_{page,position}\right)$$

$$\varepsilon_{subject,page,position} \sim N\left(0, \sigma_{subject,page,position}\right)$$

The results of the model analysis for the viewing duration, product choice, and decision time are listed in Table 2, Table 3, and Table 4 respectively with the corresponding standard errors in parentheses. In each table, we differentiated the model analysis with respect to the underlying product parameter adjustments used for determining the effects. These include: (1) all product parameters, (2) parameters directly controlled by the seller, and (3) parameters not or only indirectly controlled by the seller.

The demographic variables control for any possible differences between the test subject groups A and B. Age is centred on the sample mean to facilitate interpretation. As the results in Table 2 reveal, age does not significantly correlate with viewing duration, albeit test subjects with an above-average age seem to look at an AOI shorter in general. Likewise, male subjects appear to look longer than female participants. Similar results for the demographical data can be observed in Table 3 and Table 4.

Table 2 shows that the effect of an increased viewing duration due to the existence of a visual cue is significant (confirming *Hypothesis 1*). The effect has the expected sign and is statistically significant at the 10% level. However, when differentiated according to the control level, it emerges that only the parameters controlled by Amazon are significant, while the seller-controlled parameters hardly show an effect on the viewing duration.

Table 3 reveals that the visual cue's existence seems to have a notably large and significant effect on the product choice at the 5% level (confirming *Hypothesis 2*). Much

**Table 2.** Results of the model analysis for the viewing duration.

| | | Dependent variable: viewing duration | | |
|---|---|---|---|---|
| | | All product parameters | Parameters directly controlled by the seller | Parameters not or only indirectly controlled by the seller |
| Independent variables | Visual cue | 0.095* (0.058) | 0.022 (0.083) | 0.182** (0.079) |
| | Age | −0.037 (0.067) | −0.032 (0.070) | −0.034 (0.066) |
| | Gender (male) | 0.185 (0.161) | 0.223 (0.170) | 0.137 (0.157) |
| | Constant | 0.689*** (0.146) | 0.674*** (0.163) | 0.723*** (0.146) |
| | Observations | 2,081 | 1,038 | 1,043 |

*Note* *$p < 0.1$; **$p < 0.05$; ***$p < 0.01$.

**Table 3.** Results of the model analysis for the product choice.

| | | Dependent variable: product choice | | |
|---|---|---|---|---|
| | | All product parameters | Parameters directly controlled by the seller | Parameters not or only indirectly controlled by the seller |
| Independent variables | Visual cue | 0.458** (0.214) | 0.272 (0.327) | 0.604** (0.286) |
| | Viewing dur. | 0.386*** (0.024) | 0.383*** (0.034) | 0.388*** (0.035) |
| | Age | 0.0004 (0.012) | 0.003 (0.017) | −0.003 (0.017) |
| | Gender (male) | −0.285* (0.154) | −0.269 (0.220) | −0.298 (0.215) |
| | Constant | −2.702*** (0.330) | −2.782*** (0.468) | −2.610*** (0.464) |
| | Observations | 2,081 | 1,038 | 1,043 |

*Note* *$p < 0.1$; **$p < 0.05$; ***$p < 0.01$.

like the data in Table 2, the results are only significant for the Amazon-controlled parameters. Overall, we find that viewing duration positively correlates with product choice and is highly statistically significant.

Table 4 demonstrates that visual cues reduce the decision time of customers (confirming *Hypothesis 3*). However, in contrast to the previous results, visual cues show a

**Table 4.** Results of the model analysis for the decision time.

| | | Dependent variable: decision time | | |
|---|---|---|---|---|
| | | All product parameters | Parameters directly controlled by the seller | Parameters not or only indirectly controlled by the seller |
| Independent variables | Visual cue | −0.048* (0.026) | −0.068* (0.036) | −0.003 (0.034) |
| | Age | −0.007 (0.025) | −0.007 (0.014) | −0.008 (0.014) |
| | Gender (male) | 0.211 (0.308) | 0.271 (0.178) | 0.151 (0.171) |
| | Constant | 2.872*** (0.649) | 2.837*** (0.374) | 2.882*** (0.360) |
| | Observations | 359 | 179 | 180 |

*Note* $*p < 0.1; **p < 0.05; ***p < 0.01$.

significant effect only for the seller-controlled parameters. The relatively small reduction can be explained by the general setting of the experiment. There was no time limit and test subjects were asked to take time for their decision in order to emulate a realistic shopping situation. Thus, even if a cue reduced search costs, and thereby the evaluation time for a certain product, test subjects might have been inclined to "re-invest" the saved time into the closer examination of the remaining products.

# 6 Conclusion

## 6.1 Summary

In today's world, e-commerce and online shopping are on a continuous rise. Many consumers prefer online shopping due to convenience and a much greater range of available products [59, 60]. However, on online shopping platforms like Amazon, this variety and the fierce competition in price, performance, and visual saliency lead to an information overload making it difficult for sellers to stand out in the marketplace and compete for the attention of potential customers. The goal of this study was to examine how the existence of elements like the "Bestseller" badge or discount vouchers in product offers influence customer behaviour on the Amazon marketplace. For answering the research question, an eye-tracking study was conducted to investigate the effects of modified visual cues on the viewing duration, product choice, and decision time of customers. We drew from the attentional drift diffusion model and the theory of rational inattention to accumulate empirical evidence for the theoretical linkages between viewing duration, product choice, and decision time. Based on this, we developed three hypotheses regarding the effects of visual cues. For estimating the magnitude of their influence, we used a lognormal mixed-effects model and logistic regression. Hypothesis 1 could be confirmed on a superordinate level, i.e. the existence of visual cues can indeed lead to an increased viewing duration of a product on Amazon search pages. Visual cues also positively influence the product choice of customers, which proves Hypothesis 2. Hypothesis 3 was also verified, but visual cues had a minimal effect on the decision time of customers.

## 6.2 Managerial Implications

From a managerial perspective, it is ex-ante not always clear if visual cues can actually lead to an improved outcome, which is because there might be some reservations regarding their application. As pointed out in the introduction section, there is a risk of overdoing. For instance, a new element added to a user interface always represents an additional visual component that may be distracting if that component is not helpful to the customer. However, our empirical design allows drawing the important conclusion that visual cues like badges do indeed change customer behaviour in a way that aligns with sellers' business objectives (e.g. increased attention and higher choice propensity).

Our study provides the basis for several managerial implications for e-commerce platforms in general and for Amazon marketplace sellers in particular. The results demonstrate that e-commerce platform owners can profit by marking popular products on their websites because this can significantly increase both the viewing duration of those products and the probability that customers click on them for closer examination. From the perspective of Amazon marketplace sellers, achieving this goal is a more challenging task because the results clearly show that the most effective tools are the parameters controlled by the platform owner, i.e. Amazon. In our experiment, the visual cues the seller is able to directly control were found to be mostly statistically insignificant regarding the postulated hypotheses. Hence, the results indicate that sellers should, for instance, thoroughly examine whether to offer a price discount because the viewing duration and product choice might be hardly influenced by it. In this regard, sellers should pay more attention to product ratings, which can be expected to have a greater impact. Although product ratings are not directly controlled by the seller, s/he can take precautions to ensure that avoidable service failures do not occur and ask satisfied customers for rating their offers.

## 6.3 Limitations and Future Research Directions

While conducting the eye-tracking experiment, besides eye-tracking data we also recorded the mouse clicks of participants. It should be noted that the mouse clicks primarily served as a task goal for the test subjects and cannot be interpreted as purchase probability. It is conceivable that mouse clicks on individual product offers on Amazon search pages only indicate an increased interest of online shoppers. In real-world scenarios, a potential customer would probably open multiple product pages in Internet browser tabs to compare them in greater detail. The visual examination of product pages was not part of this study and can be addressed by future research.

Furthermore, we only found a significant gender effect regarding the product choice, which might be due to the female to male ratio being 1/3 in the examined dataset. As previous research suggests, males are characterised by lower visual attention in the online shopping context [61]. Hence, it might be possible that a more balanced dataset with an increased sample size reveals significant gender differences regarding the other dependent variables as well.

Finally, we did not seek to identify the isolated effect of individual visual cues, which could be a promising avenue for further research. Investigating the isolated effects of visual cues in more product categories and scenarios could allow the deducing of more detailed implications. For instance, it can be assumed that offering discount vouchers might be a more effective tool for drawing attention to high-priced products. The same can be expected for the "Bestseller" badge, which could be more influential when attached to expensive products due to the facilitation of the purchase decision-making process. Meaning that potential customers could be more inclined to trust the "swarm intelligence" reflected in the "Bestseller" badge when deciding between similarly priced expensive options.

# References

1. Huang, Y., Chai, Y., Liu, Y., Shen, J.: Architecture of next-generation e-commerce platform. Tinshhua Sci. Technol. **24**, 18–29 (2019)
2. Statista: Warum kaufen Sie manchmal lieber im Internet als im Geschäft? (2017), https://de.statista.com/statistik/daten/studie/219677/umfrage/gruende-fuer-online-shopping/
3. Guo, K., Wang, H., Song, Y., Du, Z.: The effect of online reviews on e-tailers' pricing in a dual-channel market with competition. Int. J. Mach. Learn. Cybern. **9**(1), 63–73 (2015). https://doi.org/10.1007/s13042-015-0346-5
4. Wang, Q., Yang, S., Liu, M., Cao, Z., Ma, Q.: An eye-tracking study of website complexity from cognitive load perspective. Decis. Support Syst. **62**, 1–10 (2014)
5. Dennis, A.R., Yuan, L., Feng, X., Webb, E., Hsieh, C.J.: Digital nudging: numeric and semantic priming in E-commerce. J. Inf. Mange. Sci. **37**, 39–65 (2020)
6. Palcu, J., Sudkamp, J., Florack, A.: Judgments at Gaze value: gaze cuing in banner advertisements, its effect on attention allocation and product judgments. Front. Psychol. **8**, 1664–1078 (2017)
7. Pieters, R., Wedel, M., Zhang, J.: Optimal feature advertising design under competitive clutter. Manag. Sci. **53**, 1815–1828 (2007)
8. Margarida Barreto, A.: Do users look at banner ads on Facebook? J. Res. Interact. Mark. **7**, 119–139 (2013)
9. European E-Commerce Report: European Ecommerce Report 2019 (2019)
10. Koch, L.: Amazon Dominates EU-5 Ecommerce Market (2019). https://www.emarketer.com/content/amazon-dominates-eu-5-ecommerce-market
11. Amazon Seller Report: The State of the Amazon Seller 2020 (2020)
12. Danziger, P.N.: Forbes: thinking of selling on Amazon marketplace? Here are the pros and cons (2018), https://www.forbes.com/sites/pamdanziger/2018/04/27/pros-and-cons-of-amazon-marketplace-for-small-and-mid-sized-businesses/
13. Djurica, D., Figl, K.: The effect of digital nudging techniques on customers' product choice and attitudes towards E-commerce sites. In: Twenty-Third Americas Conference on Information Systems (2017)
14. Soto-Acosta, P., Jose Molina-Castillo, F., Lopez-Nicolas, C., Colomo-Palacios, R.: The effect of information overload and disorganisation on intention to purchase online. Online Inf. Rev. **38**, 543–561 (2014)
15. Li, C.-Y.: Why do online consumers experience information overload? An extension of communication theory. J. Inf. Sci. **43**, 835–851 (2017)

16. Amazon: Sponsored Brands (2022). https://advertising.amazon.com/en-us/solutions/products/sponsored-brands?ref_=a20m_us_hnav_lng_en_us

17. Focus: Gibt keinen Grund "Amazon's Choice" zu vertrauen: Was wirklich hinter dem Siegel steckt (2020). https://www.focus.de/finanzen/news/online-shopping-es-gibt-keinen-grund-amazon-s-choice-zu-vertrauen-das-steckt-wirklich-hinter-dem-siegel_id_11596405.html

18. Amazon: Über Amazon Verkaufsrang (2022). https://www.amazon.de/gp/help/customer/display.html?nodeId=527588

19. Jacob, R.J.K., Karn, K.S.: Eye Tracking in Human-Computer Interaction and Usability Research. In: The Mind's Eye: Cognitive and Applied Aspects of Eye Movement Research, pp. 573–605. Elsevier (2003)

20. Duchowski, A.T.: A breadth-first survey of eye-tracking applications. Behav. Res. Methods Instrum. Comput. J. Psychon. Soc. Inc. **34**, 455–470 (2002)

21. Duchowski, A.T.: Eye Tracking Methodology. Springer, Cham (2017). https://doi.org/10.1007/978-1-84628-609-4

22. Ehmke, C., Wilson, S.: Identifying Web Usability Problems from Eye-Tracking Data British HCI Conference. University of Lancaster, UK (2007)

23. Lakshmi Pavani, M., Bhanu Prakash, A.V., Shwetha Koushik, M.S., Amudha, J., Jyotsna, C.: Navigation through eye-tracking for human–computer interface. In: Satapathy, S.C., Joshi, A. (eds.) Information and Communication Technology for Intelligent Systems. SIST, vol. 107, pp. 575–586. Springer, Singapore (2019). https://doi.org/10.1007/978-981-13-1747-7_56

24. Jankowski, J., Wątróbski, J., Witkowska, K., Ziemba, P.: Eye tracking based experimental evaluation of the parameters of online content affecting the web user behaviour. In: Nermend, K., Łatuszyńska, M. (eds.) Selected Issues in Experimental Economics. SPBE, pp. 311–332. Springer, Cham (2016). https://doi.org/10.1007/978-3-319-28419-4_20

25. Adelmeyer, M., Beinke, J., Walterbusch, M., Gameiro, R., König, P., Teuteberg, F.: Eye-tracking zur Untersuchung von Vertrauenssignalen auf Websiten von Cloud Computing-Anbietern. In: Mayr, H., Pinzger, M. (eds.) INFORMATIK 2016: Lecture Notes in Informatics (LNI), pp. 883–896. Bonn (2016)

26. Bergstrom, J.R., Schall, A.J. (eds.): Eye tracking in User Experience Design. Elsevier, Amsterdam (2014)

27. Pan, B., Hembrooke, H.A., Gay, G.K., Granka, L.A., Feusner, M.K., Newman, J.K.: The determinants of web page viewing behavior. In: Duchowski, A.T., Vertegaal, R. (eds.) Proceedings of the Eye Tracking Research & Applications Symposium on Eye Tracking Research & Applications - ETRA'2004, pp. 147–154. ACM Press, New York, New York, USA (2004)

28. Roth, S.P., Tuch, A.N., Mekler, E.D., Bargas-Avila, J.A., Opwis, K.: Location matters, especially for non-salient features–an eye-tracking study on the effects of web object placement on different types of websites. Int. J. Hum. Comput. Stud. **71**, 228–235 (2013)

29. Clark, J.W., Stephane, A.L.: Affordable eye tracking for informed web design. In: Marcus, A., Wang, W. (eds.) DUXU 2018. LNCS, vol. 10918, pp. 346–355. Springer, Cham (2018). https://doi.org/10.1007/978-3-319-91797-9_24

30. Sari, J.N., Nugroho, L.E., Insap Santosa, P., Ferdiana, R.: Modeling of consumer interest on e-commerce products using eye tracking methods. In: Proceedings of the International HCI and UX Conference in Indonesia on – CHIuXiD 2015, vol. 700, pp. 147–157. ACM Press, New York (2015)

31. Wątróbski, J., Jankowski, J., Karczmarczyk, A., Ziemba, P.: Integration of eye-tracking based studies into e-commerce websites evaluation process with eQual and TOPSIS methods. In: Wrycza, S., Maślankowski, J. (eds.) SIGSAND/PLAIS 2017. LNBIP, vol. 300, pp. 56–80. Springer, Cham (2017). https://doi.org/10.1007/978-3-319-66996-0_5

32. Schmutz, P., Roth, S.P., Seckler, M., Opwis, K.: Designing product listing pages—effects on sales and users' cognitive workload. Int. J. Hum Comput Stud. **68**, 423–431 (2010)

33. Wang, Q., Xu, Z., Cui, X., Wang, L., Ouyang, C.: Does a big Duchenne smile really matter on e-commerce websites? An eye-tracking study in China. Electr. Comm. Res. **17**, 609–626 (2017)
34. Lee, J., Ahn, J.-H.: Attention to banner ads and their effectiveness: an eye-tracking approach. Int. J. Electr. Comm. **17**, 119–137 (2012)
35. Fei, M., Tan, H., Peng, X., Wang, Q., Wang, L.: Promoting or attenuating? An eye-tracking study on the role of social cues in e-commerce livestreaming. Decis. Support Syst. **142**, 113466 (2021)
36. Yang, S.-F.: An eye-tracking study of the Elaboration Likelihood Model in online shopping. Electron. Commer. Res. Appl. **14**, 233–240 (2015)
37. Vieira, V.: Stimuli–organism-response framework: a meta-analytic review in the store environment. J. Bus. Res. **66**, 1420–1426 (2013)
38. Just, M.A., Carpenter, P.A.: Eye fixations and cognitive processes. Congn. Pschol. **8**, 441–480 (1976)
39. Eckstein, M.K., Guerra-Carrillo, B., Miller Singley, A.T., Bunge, S.A.: Beyond eye gaze: what else can eyetracking reveal about cognition and cognitive development? Dev. Cogn. Neurosci. **25**, 69–91 (2017)
40. Orquin, J.L., Perkovic, S., Grunert, K.G.: Visual biases in decision making. Appl. Econ. Perspect. Pol. **40**, 523–537 (2018)
41. Ahn, J.-H., Bae, Y.-S., Ju, J., Oh, W.: Attention adjustment, renewal, and equilibrium seeking in online search: an eye-tracking approach. J. Manage. Inf. Syst. **35**, 1218–1250 (2018)
42. Schotter, E.R., Berry, R.W., McKenzie, C.R.M., Rayner, K.: Gaze bias: selective encoding and liking effects. Vis. Cogn. **18**, 1113–1132 (2010)
43. Farnsworth, B.: 10 Most Used Eye Tracking Metrics and Terms (2020). https://imotions.com/blog/10-terms-metrics-eye-tracking/#first
44. Holmqvist, K., Nyström, M., Andersson, R., Dewhurst, R., Halszka, J., van de Weijer, J.: Eye Tracking a Comprehensive Guide to Methods and Measures. Oxford University Press, Oxford (2011)
45. Poole, A., Ball, L.J.: Eye Tracking in Human-Computer Interaction and Usability Research: Current Status and Future Prospects. Psychology Department, Lancaster University, UK (2005)
46. Tavares, G., Perona, P., Rangel, A.: The attentional drift diffusion model of simple perceptual decision-making. Front. Neurosci. **11**, 468 (2017)
47. Peschel, A.O., Orquin, J.L., Mueller Loose, S.: Increasing consumers' attention capture and food choice through bottom-up effects. Appetite **132**, 1–7 (2019)
48. Friedrich, T., Figl, K.: Consumers' perceptions of different scarcity cues on E-commerce websites. In: Proceedings of the 39th International Conference on Information Systems (ICIS), vol. 9 (2018)
49. Krajbich, I., Lu, D., Camerer, C., Rangel, A.: The Attentional Drift-Diffusion Model Extends to Simple Purchasing Decisions. Front. Phycol. **3**, 1664–1078 (2012)
50. Hauser, J.R., Wernerfelt, B.: An evaluation cost model of consideration sets. J. Consum. Res. **16**, 393 (1990)
51. Roberts, J.H., Lattin, J.M.: Development and testing of a model of consideration set composition. J. Mark. Res. **28**, 429–440 (1991)
52. Los Santos, B. de, Hortaçsu, A., Wildenbeest, M.R.: Testing models of consumer search using data on web browsing and purchasing behavior. Am. Econ. Rev. **102**, 2955–2980 (2012)
53. Sims, C.A.: Implications of rational inattention. J. Monet. Econ. **50**, 665–690 (2003)
54. Amazon: Anforderungen an Produktbilder (2022). https://sellercentral.amazon.de/gp/help/external/G1881?language=de_DE&ref=efph_
55. Amazon: Sponsored Products (2022). https://advertising.amazon.com/en-us/solutions/products/sponsored-products

56. Amazon: Amazon FBA: Save time and help grow your business with FBA (2022), https://sell.amazon.com/fulfillment-by-amazon.html

57. Zimoch, M., Pryss, R., Schobel, J., Reichert, M.: Eye Tracking experiments on process model comprehension: lessons learned. In: Reinhartz-Berger, I., Gulden, J., Nurcan, S., Guédria, W., Bera, P. (eds.) BPMDS/EMMSAD -2017. LNBIP, vol. 287, pp. 153–168. Springer, Cham (2017). https://doi.org/10.1007/978-3-319-59466-8_10

58. Bera, P., Soffer, P., Parsons, J.: Using eye tracking to expose cognitive processes in understanding conceptual models. Manag. Inf. Syst. Q. **43**, 1105–1126 (2019)

59. Dolfen, P., et al.: Assessing the Gains from E-Commerce. National Bureau of Economic Research, Cambridge (2019)

60. Ganesh, J., Reynolds, K.E., Luckett, M., Pomirleanu, N.: Online shopper motivations, and e-store attributes: an examination of online patronage behavior and shopper typologies. J. Retail. **86**, 106–115 (2010)

61. Hwang, Y.M., Lee, K.C.: Using an eye-tracking approach to explore gender differences in visual attention and shopping attitudes in an online shopping environment. Int. J. Hum.-Comput. Interact. **34**, 15–24 (2018)

# The Use of Mixed-Reality Sport Platforms in Social Media Sentiment Analysis during COVID-19

László Dömök$^{(\boxtimes)}$ and Szabina Fodor

Corvinus University of Budapest, Fővám tér 13-15., Budapest 1093, Hungary
laszlo.domok@stud.uni-corvinus.hu, szabina.fodor@uni-corvinus.hu

**Abstract.** Sport and physical activity are very closely related to people's health. The COVID-19 pandemic has made everyone aware of the importance of maintaining regular physical activity. The lockdowns and mandatory social distancing experienced during the epidemic underlined the importance of new sports platforms that bring traditional sports, such as cycling, to the virtual world. This work focuses on the ZWIFT cycling application as an exemplary mixed-reality sport platform. Sentiment analysis (or opinion mining) aims to explore the emotions behind the opinions expressed in texts on different topics. We used sentiment analysis of social media platforms (Twitter and Reddit) to provide valuable information on the culture surrounding mixed reality sports platforms.

**Keywords:** Mixed reality · Virtual sports · Sentiment analysis · Social network · COVID-19

## 1 Introduction

Physical activity plays an important role in preventing and treating many chronic diseases, which are responsible for almost 70% of deaths worldwide [1]. Amateur participation sports such as running, triathlon, and cycling are severely affected by the COVID-19 pandemic, as these sports depend heavily on mass physical participation. In 2020, most areas of the sports industry, including competitive and recreational sport systems and players, related sports markets, sports tourism, and sports media sectors, experienced significant declines [2–4]. In the world of sport, it should be said that since the introduction of restrictions, there have been major changes in physical activity and sporting habits of populations around the world [5]. Stockwell et al. looked at the results of 64 representative surveys conducted in countries around the world and concluded that during the first wave of the COVID-19 pandemic, the population almost everywhere experienced a decrease in physical activity and an increase in sedentary time. Lower-intensity activities (e.g. walking) are more prevalent in the changed physical activity patterns than in the pre-restriction period. Based on demographic and sociocultural characteristics, significant differences were found in terms of age, suggesting that younger age groups (under 30 years) were more likely to continue to find opportunities for physical activity,

Ē. Nazaruka et al. (Eds.): BIR 2022, LNBIP 462, pp. 175–188, 2022.
https://doi.org/10.1007/978-3-031-16947-2_12

compared to older age groups (31–65 and >65), who significantly reduced their time spent playing sport by more than 50% [6].

There were also marked changes in the places where the sport was played during the COVID-19 pandemic. On the one hand, restrictions first affected community venues such as sports, fitness and health centres, and sports clubs, and on the other hand, fear of the virus and the appeal kept a significant proportion of people away from their usual sporting venues [7, 8]. At the same time, many global sporting competitions have been postponed or cancelled, opening up the possibility of a virtual platform as an alternative to professional sporting competitions [9].

The emergence of the epidemic and the restrictions imposed have increased the opportunities and forms of home sport, with a significant increase in the frequency of individual sport and physical activity in nature [7]. The increase in individual outdoor sporting activities is illustrated by the information recorded by various so-called 'smart devices' on people's physical activity. During the period of restrictions, there was a significant increase in the proportion of people who participated in online training and the time they spent doing so. During this period, several improvements were made, covering software, the tools that could be used, and online sports training materials. These have created an environment of leisure activity at home and in many cases even conditions for competition (e.g., cycling) [10]. The boundaries between real and virtual sports are increasingly blurred, with the emergence of a mixed sports model that combines the physical exertion of sports with the virtual representation, adding a new dimension. It has to be said that home-based training is presumably an easier option for those who already have the necessary equipment, space, and competence to carry out self-guided training and physical movements. Furthermore, adherence to home exercises requires a complex and high level of intrinsic motivation [11].

Several studies [12, 13] investigated the impact of digital solutions on work and personal life, but the COVID-19 pandemic has shown that digital alternatives to the sport are still significantly missing from our daily lives, even though the sport is, in many cases, an integral part of life. There are some new types of applications that allow users to perform physical activity and virtual mediated interaction while being embedded in a virtual world. One of the most popular mixed-reality sport platforms is ZWIFT [9], which offers the possibility to link the bike to the ZWIFT application using a smart trainer and display the real performance in a virtual world and offers a digital exercising alternative to users.

In this work, we apply sentiment analysis of social media platforms to provide valuable insight into the culture surrounding the mixed reality sports platform. Through an analysis of Twitter and Reddit posts related to the ZWIFT application, we sought to understand the impact of the restrictions caused by the COVID-19 virus outbreak on the application's users. Did they feel that they had the opportunity to exercise and train with others despite their circumstances? Can we see a change in the number of comments during the lockdowns, i.e. did more people noticeably and intensively use this platform to exercise? Is the stress of fear and vulnerability reflected in the comments posted on specific social media channels?

The remainder of this paper is structured as follows. In Sect. 2, we give an overview of sports digitalization and ZWIFT as an exemplary mixed reality sports application.

Subsequently, we present the design of our investigation in which we used data retrieved between January 1, 2018, and December 31, 2021, from the Twitter and Reddit social media platforms in Sect. 3. The results of the usage of the ZWIFT platform are presented in Sect. 4. This paper ends with a conclusion and future outlook in Sect. 5.

## 2 Digitalisation on Sport

Sport is today a highly valuable industry worldwide and can be defined in business terms in several different areas: for example, as a live sport, recreation, fitness, and entertainment (sports broadcasting). All of these areas have been significantly transformed by digital technology over the last few years, including the fact that it is now available to the amateur runner, something that was previously unimaginable even in elite sports. Data, sensors, digital services, and channels, social networks have overwhelmed tradition, and we have not even mentioned virtual and augmented reality. These technologies not only provide new tools for training for professional and amateur athletes, but with their proliferation, new sports and business models are emerging.

Nowadays, even amateur athletes have many possibilities to track and analyse their performance, especially with the help of wearable sensor devices or even smartphone apps. The market for such technologies is estimated to reach $12 billion by 2022 [14], with more and more big sports brands trying to grab a slice of the market, seeing the growing demand. Another important technology is virtual reality, where virtual running and cycling apps simulate and visualise the realistic completion of a course on training machines, such as the Zwift (https://www.zwift.com), and Peloton (https://www.onepeloton.com/), Real Grand Tours (RGT; https://www.rgtcycling.com/), Rouzy (https://rouvy.com/). This new form of sport platforms transfers cycling into a virtual world and allows users to socialize, exercise, or compete with each other [15]. These training systems offer several benefits, such as allowing users to train and compete with each other regardless of weather and geographical conditions. It also allows accurate and repeatable control of the characteristics of the virtual environment [16].

Currently, Zwift is the market leader in simulated cycling races [9], with more than 2.5 million registered app users in 190 countries [17] and more than 30 000 users who can both cycle [18] and compete with each other, as in addition to traditional rides in the virtual world, users can also compete with other cyclists (see Fig. 1), access training plans created by professional coaches and even train in groups. Zwift's community is constantly growing, racing teams are being formed and virtual races are monitored by a third party (Zwiftpower; https://zwiftpower.com/) to ensure transparency. During the COVID-19 outbreak, several new additions were available to the ZWIFT world of routes. The Richmond and Innsbruck routes are replicas of the 2015 and 2018 World Road Championships courses, while Watopia is a fictional world created based on locations somewhere in the Solomon Islands. The New York one is mainly based on routes in Central Park. ZWIFT combines experiential cycling with opportunities for structured training, and also allows for community cycling. Furthermore, gamification elements [19] have been added to the platform, such as avatar individualisation and so-called powerups (e.g. invisibility for a few seconds).

**Fig. 1.** ZWIFT mixed-reality sport platform, allows users to ride their bikes in a virtual world. Using the smart-trainer allows you to cycle with other cyclists in a virtual environment (Source: https://www.bikeradar.com/features/ZWIFT-your-complete-guide/)

## 3   Material and Methods

Sentiment analysis has become a popular tool in both business and academia [20, 21]. Since the internet has made large amounts of data available, sentiment analysis is increasingly used in marketing, social sciences and even politics [22]. By definition, sentiment analysis, or opinion mining, deals with the study of emotions and attitudes expressed in a text [23]. Sentiment analysis methods can be classified into several categories, both vertically and horizontally. Vertically, we can talk about aspect-based, sentence-based, and document-based analysis [24, 25]. Horizontally, different techniques can be used, two techniques are important to mention, such as lexicon (or dictionary) based and machine learning. The dictionary-based approach is based on assigning an emotional charge to the words in a given text using an emotion dictionary. Machine learning methods, on the other hand, perform text classification through supervised or unsupervised machine learning [26].

Nowadays, it is increasingly popular to use data and texts accumulated on social media as data sources [25]. Social media is already used by billions of people around the world. Almost half a million posts are published on Reddit every minute, 63% of the world's internet users use Facebook and 23% use Twitter every week, so there is an almost limitless amount of information available through them, which can provide a good basis for analysis [27]. Twitter allows to users post short - 280 characters long, called tweets [25]. Reddit is a social online forum where users can share their thoughts with a limit of 40,000 characters [26]. Both social platforms are popular with internet users and offer an accessible and searchable database for academic research and business.

We summarize how our methodology works in Fig. 2. In the first step, the posts on social media were downloaded from Twitter and Reddit from 1 January 2018 to 31 December 2021. The number of rows that have been downloaded from Twitter is 617,488 and tweets, from Reddit is 168,666 posts. Next, the data was pre-processed to remove redundant data, inconsistent data and noisy data. In the pre-processing, duplicates were removed from both social media datasets and the data was formatted for data analysis. After these steps, 613,003 tweets on Twitter and 168,085 posts on Reddit have remained. A two-step sentiment analysis was performed (see Fig. 2). VADER model was used to obtain the sentiment scores of the comments and categorise them into "positive", "neural" and "negative". In the second step of the sentiment analysis, to see the emotions behind the categories of positive and negative posts, the NRCLex package was used. An unsupervised Latent Dirichlet Allocation (LDA) topic modeling [28] was performed built to identify commonly discussed topics based on neutral posts.

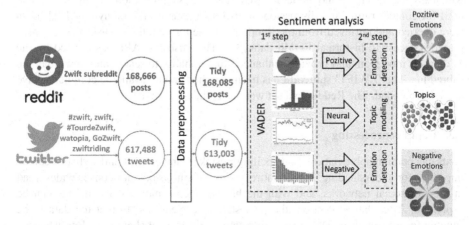

**Fig. 2.** An overview of our research methodology (Source: prepared by the authors)

### 3.1 Data Collection - Twitter Data

The Twitter site has an average of 436 million monthly active users in over 100 countries in January 2022, providing researchers with data conveniently accessible using Twitter APIs [27]. Twitter has given topics a 'hashtag' (the hashtag is '#'), which works like tags, referring to a topic. Users can tag their newly written post with such a tag, or they can easily search for specific topics by clicking on the tags. In this way, all posts on a topic can be listed and reviewed very quickly. The data shows that most Twitter users tweet only once a month on average, and 10% of Twitter users are responsible for 92% of all tweets by all US users. This means that most users visit Twitter to keep up with the latest news, not to talk to and follow communities of their interest [29].

The data was downloaded using the official Twitter API v2 with academic research access, making millions of public tweets available. This Twitter API v2 allows for the download of complete and unbiased data, and enables more accurate data retrieval and download for analysis of public conversations, free of charge for eligible researchers, up to 10 000 000 tweets a month [30]. The following search terms were used for our queries: '#ZWIFT', 'ZWIFT', '#TourdeZWIFT', 'watopia', 'GoZWIFT', 'ZWIFTriding'. These keywords refer to the Zwift platform, for example, 'TourdeZwift' is a popular multi-day cycling event on the platform and 'watopia' is a virtual location. All available tweets have been searched, and retweets and promoted tweets have been excluded from the query. The data set was stored in csv format.

### 3.2  Data Collection - Reddit Data

Reddit is a slightly different social portal from Twitter, as the focus is on the community rather than the individual user. Users can post text messages, images, or video content and engage in discussions with other users in predefined groups, called subreddits, which anyone can create and moderate. Reddit is a news aggregation website that is also used for content rating and discussion, it also maintains users' anonymity [31]. Data were queried and retrieved from the Reddit database through the Pushshift API, which provides access to historical Reddit data [32]. The Pushshift API was created by the /r/ datasets mod team and provides enhanced functionality to search and download the subreddits. The 'ZWIFT' subreddit has been searched and all available comments have been downloaded. The Reddit data set was stored in csv format.

### 3.3  Data Preprocessing

Texts can contain abbreviations and spelling mistakes, and context and stylistics can be important for understanding. Furthermore, the use of emojis or emoticons is widespread in modern social networks, and with the help of these emotions and moods can be expressed. After data collection, the next step is to clean and format the data to be suitable for text analysis. Data processing steps: urls, special characters, English stop-words, numbers, and unnecessary spaces have been deleted. Furthermore, the text was transformed through the use of tokenization and lemmatization.

### 3.4  Sentiment Analysis

In the first step of our research, we used VADER, Valence Aware Dictionary and sEntiment Reasoner [33] to determine the polarity and intensity of the post. VADER is a sentiment lexicon and rule-based sentiment analysis tool, which is exclusively standardized to the sentiments manifested in social media. The built in lexicon was compiled by 10 independent, trained experts, evaluating more than 9000 words or phrases [34, 35]. As a first step, VADER was applied to obtain sentiment scores for pre-processed English textual data. Furthermore, following the classification method recommended by the authors of the tool, sentiment scores were mapped into three categories: positive, negative and neutral.

In the second step of our analysis, we used the nrc [36] algorithm to discover the emotions underlying categories of both negative and positive posts. This algorithm uses an emotion dictionary to score each post based on two sentiments (positive or negative) and eight emotions (anger, fear, anticipation, trust, surprise, sadness, joy, and disgust). This analysis helps to understand the emotions carried by posts categorised as negative and positive. Although the nrc algorithm also categorizes posts positively and negatively, the literature suggests that these categories are not as precise as in the case of VADER. Thus, we did not use the results of the nrc algorithm for polarity, we used only VADER to explore the polarity of our posts and the nrc algorithm to explore the emotions behind negative and positive posts.

Posts categorised as neutral by VADER can be a good starting point for exploring the topics discussed on Twitter and Reddit. We used a topic model approach in the second step of our research, which is an unsupervised machine learning method. One of the most popular of these techniques is the latent dirichlet allocation (LDA) model, which is a probabilistic model based on a three-level hierarchical Bayesian model for expressing a corpus. In our analysis of tweets and comments on Reddit over the 4 years, we identified the most common topics discussed on social media.

## 4   Results and Discussion

The change in popularity of ZWIFT has been examined on social media during the period under review. Tweets and Reddit posts were collected between January 2018 and December 2021. Figure 3 shows the frequency of searches for terms on Twitter and the frequency of posts on the 'ZWIFT' subreddit.

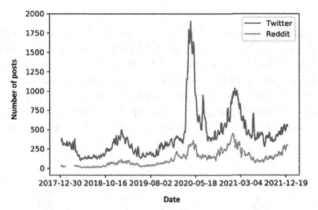

**Fig. 3.** Distribution of ZWIFT-related posts on Twitter and Reddit between 1 January 2018 and 31 December 2021 (Source: prepared by the authors)

Figure 3 illustrates how, with the outbreak of the coronavirus epidemic, the number of posts on ZWIFT Twitter skyrocketed in the first half of 2020. This can be partly explained by the cancellation of traditional cycling races due to the epidemic, which ZWIFT replaced with the '2020 Tour de Zwift', an online race that was a mass event open to all, with 119,700 riders participating and over 15,500 h of content watched by race fans on Tour de Zwift broadcasts. This spike is also seen on Reddit, but to a lesser extent. This difference can be partly explained by differences in user habits between the two platforms. While Twitter users focus on breaking news and results, Reddit users tend to discuss a single topic. It is worth noting that, regardless of the pandemic, there is a cyclical pattern in the number of comments, with online platforms and online cycling being more popular in winter months than in summer.

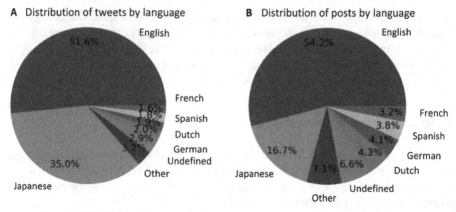

**Fig. 4.** Distribution of comments by language (Source: prepared by the authors)

Not surprisingly, more than half of the posts on the ZWIFT subreddit (panel B of Fig. 2) are in English, and on Twitter (panel A of Fig. 2) most of the posts are also in English, but about half of the remaining posts are in Japanese (see Fig. 4). This is an unexpectedly high value, given that, according to 2022 data, around 2.2% of the world's internet content is in Japanese only (https://www.statista.com/). This is in line with several studies [37, 38] on the adoption and acceptance of technology, which show that Japanese citizens are more readily accepting of advances in automation and AI-enabled products.

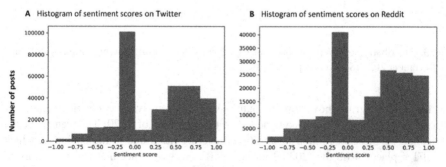

**Fig. 5.** Histogram of sentiment scores on Twitter and Reddit (Source: prepared by the authors)

Based on the VADER-Sentiment-Analysis algorithm, 1 compound score means the most extreme positive score and $-1$ compound score means the most extreme negative score. Figure 5 illustrates that both on social networks the positive sentiments are the most frequent. There are minimal differences between the histograms. The typical threshold values of VADER sentiments are: a) positive sentiment: compound score $\geq 0.05$; b) neutral sentiment: (compound score $> -0.05$) and (compound score $< 0.05$) and c) negative sentiment: compound score $\leq -0.05$. Based on this classification, Fig. 5 shows the distribution of the positive, neutral and negative posts on Twitter and Reddit. The sentiment of most words is neural; they also tend to be more positive than negative; the proportion of positive posts is around 60% on both platforms, but the negative sentiment is slightly more common on Reddit (Twitter: 13% vs Reddit: 18%).

We also examined the change in sentiment values over time. Figure 6 shows that the compound scores were throughout the positive range in the period on both platforms. In early 2020, Twitter's score fell slightly but did not hit a negative range. On Reddit, there was no notable fluctuation in sentiment scores during the coronavirus epidemic.

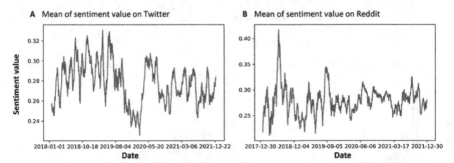

**Fig. 6.** Mean sentiment values on social media platforms between 1 January 2018 and 31 December 2021 (Source: prepared by the authors)

The decline in compound sentiment scores on Twitter in early 2020 was caused by a decline in positive posts (see Fig. 7). Furthermore, the number of positive tweets on Twitter decreased slightly during the coronavirus crisis, while the number of negative tweets did not change. On Reddit, there was no change in positive and negative posts during coronavirus.

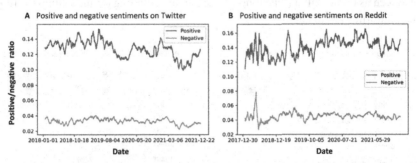

**Fig. 7.** Positive and negative sentiment values on social media platforms between 1 January 2018 and 31 December 2021 (Source: prepared by the authors).

We looked at the 20 most frequent words in positive and negative posts (see Fig. 8). The two most frequent words in positive posts 'like' and 'good' are in the same order on both platforms and 'great' and 'thanks' are among the most frequent expressions on both Twitter and Reddit. The same is true for words in negative posts, the two most frequent expressions are 'pain' and 'problem' on both platforms.

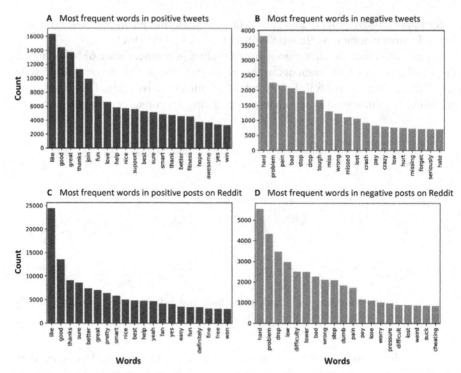

**Fig. 8.** Most frequent words on positive and negative posts on Twitter and Reddit based on VADER sentiment classification

It is worth noting that, for both negative and positive sentiment posts, there are clear themes such as competition ('cheating', 'lose', 'win'), training ('pain', 'fitness', 'fun'), digital platform ('pay', 'smart').

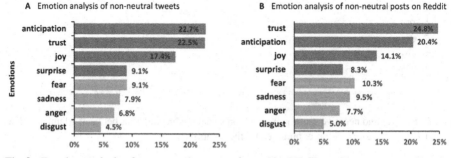

**Fig. 9.** Emotion analysis of non-neutral posts performed by NRCLex (Source: prepared by the authors).

The second step of the sentiment analysis focuses on searching for emotions in non-neutral posts. Among the eight basic emotions, 'trust' and 'joy' were the prominent positive emotions observed, while 'fear', and 'sadness' were the top negative emotions (see Fig. 9). It's worth noting that no substantial differences can be recognised between the social media platforms we studied.

To explore what the user is concerned about on Twitter and Reddit regarding the usage of ZWIFT, we applied LDA to our neutral classified corpus. For a comprehensive better representation of the content, it is necessary to find an appropriate topic number. By using topic numbers k ranging from 3 to 10, we initialized the LDA models and calculated the model coherence. Finally, we chose 4 as the topic number: the model has no intersections among topics, summarizes the whole word space well, and the topics remain relatively independent. Our model output the following four themes:

Twitter:

- Topic-1: virtual cycling
- Topic-2: workout and racing
- Topic-3: virtual racing event
- Topic-4: sharing activity

Reddit:

- Topic-1: virtual cycling
- Topic-2: workout and racing
- Topic-3: digital tools
- Topic-4: users' performance

The first topic of both Twitter and Reddit posts focuses on virtual cycling and related devices. The group is defined by the terms 'bike', 'trainer', 'app' and 'wheel'. While the words that define the group in the Twitter posts are 'update', 'working' and 'turbo', on Reddit the words include terms related to specific cycling hardware such as 'tacx', 'wahoo' and 'buy'.

The second theme on both platforms is training and racing, defined by the words 'ride', 'race' and 'route'. The second theme is defined by the words 'km', 'calories' and 'time', which refer to the competition between users on both platforms.

The third theme of Twitter posts is virtual events and races. The keywords for this theme are 'virtual', 'tour', 'stage', 'watopia' (course in Zwift), 'france' (course in Zwift), 'event', 'team'. In Reddit posts, virtual events are not so prominent, here the third theme is about digital tools. The keywords are 'app', 'tv', 'phone', 'bluetooth', 'test', 'issue'.

Twitter's fourth theme is related to users' sharing activities, the most relevant terms being 'streaming', 'podcast' and 'watch'. Reddit's fourth topic is related to users' performance and results, the most relevant terms being 'watts', 'deleted', 'kg' and 'km'.

Overall, we can say that very similar topics are discussed on the two social media platforms, with the only differences being that the Twitter community is perhaps more interested in newsworthy posts such as when tournaments are being held, what new routes are available on the virtual platform, while Reddit posts focus on experiences on specific devices.

## 5  Conclusions and Future Work

In this study, we investigated the popularity of the mixed reality sports platform during the coronavirus pandemic. We collected all available posts on the ZWIFT sports platform from Twitter and Reddit between 1 January 2018 and 31 December 2021. We found that the number of posts related to ZWIFT on both social media platforms increased in early 2020, coinciding with the emergence of the coronavirus, and the sentiment behind the posts became more positive than in the previous period. Of the highest emotions detected, 'trust' was found to be the main positive emotion, while 'fear', and 'sadness' were the top negative emotions. The topic model identified the following topics experiences with virtual cycling and tools; training and racing; virtual racing events; and activity sharing. In the posts and tweets collected, the epidemic and its associated fears do not appear in the content, which may suggest that this form of exercise may also have helped users to cope with anxiety. This may also mean that some digital practices may persist after the pandemic and that professional sports competitions in virtual space may complement traditional sports competitions beyond COVID-19. Our future plans include social network analysis, which will allow us to analyse the relationships and structures between users. Furthermore, it gives the possibility to perform a named-entity recognition to identify important entities in texts (e.g. places, dates, organizations).

**Data Accessibility.** The data for this research are available in the public repository https://git hub.com/domoklac/mixed_reality_sport_research.

# References

1. Global_Health_Estimates: Deaths by Cause, Age, Sex, by Country and by Region 2000–2016 (2018). https://www.who.int/healthinfo/global_burden_disease/estimates/en/. Accessed 10 Feb 2022
2. Skinner, J., Smith, A.C.: Introduction: sport and COVID-19: impacts and challenges for the future (volume 1). Eur. Sport Manag. Q. **21**(3), 323–332 (2021)
3. Evans, A.B., et al.: Sport in the face of the COVID-19 pandemic: towards an agenda for research in the sociology of sport. Eur. J. Sport Soc. **17**(2), 85–95 (2020)
4. Horky, T.: No sports, no spectators–no media, no money? The importance of spectators and broadcasting for professional sports during COVID-19. Soccer Soc. **22**(1–2), 96–102 (2021)
5. Stockwell, S., et al.: Changes in physical activity and sedentary behaviours from before to during the COVID-19 pandemic lockdown: a systematic review. BMJ Open Sport Exerc. Med. **7**(1), e000960 (2021)
6. Mutz, M., Gerke, M.: Sport and exercise in times of self-quarantine: how Germans changed their behaviour at the beginning of the Covid-19 pandemic. Int. Rev. Sociol. Sport **56**(3), 305–316 (2021)
7. EPRS: How coronavirus infected sport? (2021). https://www.europarl.europa.eu/RegData/etu des/BRIE/2021/659449/EPRS_BRI(2021)659449_EN.pdf
8. Grix, J., et al.: The impact of Covid-19 on sport. Int. J. Sport Policy Polit. **13**(1), 1–12 (2021)
9. McIlroy, B., et al.: Virtual training of endurance cycling–a summary of strengths, weaknesses, opportunities and threats. Front. Sports Act. Living **3**, 631101 (2021)
10. Cortis, C., et al.: Home is the new gym: exergame as a potential tool to maintain adequate fitness levels also during quarantine. Hum. Mov. **21**(4), 79–87 (2020)
11. Bachmann, C., Oesch, P., Bachmann, S.: Recommendations for improving adherence to home-based exercise: a systematic review. Physikalische Medizin Rehabilitationsmedizin Kurortmedizin **28**(01), 20–31 (2018)
12. Waizenegger, L., et al.: An affordance perspective of team collaboration and enforced working from home during COVID-19. Eur. J. Inf. Syst. **29**(4), 429–442 (2020)
13. Carillo, K., et al.: Adjusting to epidemic-induced telework: empirical insights from teleworkers in France. Eur. J. Inf. Syst. **30**(1), 69–88 (2021)

14. Ulosoy, Y.: The effect of the COVID-19 pandemic on gym, indoor and outdoors sports. J. Sport Sci. **1**(3), 17–23 (2021)
15. Westmattelmann, D., et al.: The show must go on-virtualisation of sport events during the COVID-19 pandemic. Eur. J. Inf. Syst. **30**(2), 119–136 (2021)
16. Neumann, D.L., et al.: A systematic review of the application of interactive virtual reality to sport. Virtual Reality **22**(3), 183–198 (2017). https://doi.org/10.1007/s10055-017-0320-5
17. Long, M.: Zwift Raises US\$450m in Series C Funding Round Led by KKR (2020). https://www.sportspromedia.com/
18. Schlange, E.: New "PEAK ZWIFT" Achieved: 34,940 (2020). https://zwiftinsider.com/peak-zwift-34940/
19. Hamari, J., Koivisto, J.: "Working out for likes": an empirical study on social influence in exercise gamification. Comput. Hum. Behav. **50**, 333–347 (2015)
20. Brett, E.I., et al.: A content analysis of JUUL discussions on social media: using Reddit to understand patterns and perceptions of JUUL use. Drug Alcohol Depend. **194**, 358–362 (2019)
21. Rieis, J., et al.: Breaking the news: first impressions matter on online news. In: Proceedings of the International AAAI Conference on Web and Social Media (2015)
22. Li, Y., et al.: Deep learning for remote sensing image classification: a survey. Wiley Interdisc. Rev. Data Min. Knowl. Discov. **8**(6), e1264 (2018)
23. Zhao, J., Liu, K., Xu, L.: Sentiment analysis: mining opinions, sentiments, and emotions. Comput. Linguist. **42**(3), 595–598 (2016)
24. Birjali, M., Kasri, M., Beni-Hssane, A.: A comprehensive survey on sentiment analysis: approaches, challenges and trends. Knowl.-Based Syst. **226**, 107134 (2021)
25. Yadav, A., Vishwakarma, D.K.: Sentiment analysis using deep learning architectures: a review. Artif. Intell. Rev. **53**(6), 4335–4385 (2019). https://doi.org/10.1007/s10462-019-09794-5
26. Medhat, W., Hassan, A., Korashy, H.: Sentiment analysis algorithms and applications: a survey. Ain Shams Eng. J. **5**(4), 1093–1113 (2014)
27. Statista: Social Networks Worldwide (2022). https://www.statista.com/statistics/272014/global-social-networks-ranked-by-number-of-users/
28. Blei, D.M., Ng, A.Y., Jordan, M.I.: Latent dirichlet allocation. J. Mach. Learn. Res. **3**, 993–1022 (2003)
29. Shepherd, J.: 22 Essential Twitter Statistics You Need to Know in 2022 (2022). https://thesocialshepherd.com/blog/twitter-statistics
30. Twitter_API: Developer Platform (2022). https://developer.twitter.com/en/docs/twitter-api
31. Weninger, T.: An exploration of submissions and discussions in social news: mining collective intelligence of Reddit. Soc. Netw. Anal. Min. **4**(1), 1–19 (2014). https://doi.org/10.1007/s13278-014-0173-9
32. Baumgartner, J., et al.: The Pushshift Reddit dataset. In: Proceedings of the International AAAI Conference on Web and Social Media (2020)
33. Hutto, C., Gilbert, E.: VADER: a parsimonious rule-based model for sentiment analysis of social media text. In: Proceedings of the International AAAI Conference on Web and Social Media (2014)
34. Bose, R., Dey, R.K., Roy, S., Sarddar, D.: Sentiment analysis on online product reviews. In: Tuba, M., Akashe, S., Joshi, A. (eds.) Information and Communication Technology for Sustainable Development. AISC, vol. 933, pp. 559–569. Springer, Singapore (2020). https://doi.org/10.1007/978-981-13-7166-0_56
35. Valdez, D., et al.: Social media insights into US mental health during the COVID-19 pandemic: longitudinal analysis of Twitter data. J. Med. Internet Res. **22**(12), e21418 (2020)
36. Mohammad, S., Turney, P.: Emotions evoked by common words and phrases: using mechanical turk to create an emotion lexicon. In: Proceedings of the NAACL HLT 2010 Workshop on Computational Approaches to Analysis and Generation of Emotion in Text (2010)

37. Chen, C.-F., et al.: When East meets West: understanding residents' home energy management system adoption intention and willingness to pay in Japan and the United States. Energy Res. Soc. Sci. **69**, 101616 (2020)
38. Jung, I., Lee, J.: A cross-cultural approach to the adoption of open educational resources in higher education. Br. J. Edu. Technol. **51**(1), 263–280 (2020)

# Characterizing Maturity of Digital Transformation in Organizations – A Socio-technical Framework

Fynn-Hendrik Paul[(✉)] [iD], Henning Brink [iD], and Nicole Draxler-Weber [iD]

Osnabrück University, 49069 Osnabrück, Germany
{fynn-hendrik.paul,henning.brink,nicole.draxler-weber}@uos.de

**Abstract.** Digital technologies foster organizations to rethink their business models and socio-technical structures. Thus, digital transformation (DT) has become a compelling priority on organizations' agendas. To meet the new environment, well-considered actions must be initiated and monitored at the operational and strategic levels. Therefore, it requires an understanding of fields of action and possible trajectories of DT within different organizational dimensions. For this purpose, practitioners and academics have designed numerous digital maturity models to keep track of DT progress. Still, most models reveal an incomplete picture of the holistic and socio-technical nature of DT and organizations. This motivates us to answer: *Which set of organizational dimensions and characteristics maps the holistic and socio-technical nature of DT in organizations?* With a systematic literature review and a Delphi study, our paper aims to identify and validate relevant DT-related dimensions and characteristics. The result is a socio-technical framework that serves as a pattern for (re)designing digital maturity models.

**Keywords:** Digital transformation · Digital maturity · Framework

## 1 Introduction

Digital technologies continue to grow in importance and transform the environment in which organizations operate. They cause changes in customers' requirements, in the conduct of business, and the competition and interaction between organizations [1]. To meet the new environment and ensure viability, necessary strategic and operational changes must be made within the organization [2]. In this context, the concepts of digitization, digitalization, IT-enabled organizational transformation and digital transformation (DT) are used [3–5]. While the literature on these concepts, especially on DT, focuses largely on digital technologies, these represent only one aspect of the complex phenomena [5]. Organizations represent socio-technical systems [6]. Therefore, to successfully manifest the changes throughout organizations in the long term which corresponds to a DT [3, 4], a socio-technical perspective is required [7, 8]. This means that a multitude of facets of an organization, such as business models, technology infrastructure, processes, leadership style, and culture, should be considered and aligned together [3, 9–11]. Thus, it

Ē. Nazaruka et al. (Eds.): BIR 2022, LNBIP 462, pp. 189–204, 2022.
https://doi.org/10.1007/978-3-031-16947-2_13

is crucial that managers continuously monitor progress of the organization's DT [12]. In other words, they need to keep the organization's digital maturity in sight. Digital maturity models (DMMs) help assess such progress of transformation activities in a digital context and point out a focused path throughout the transformation [13–15]. With this intention, different DMMs have been proposed in recent information systems (IS) research. However, studies show that most DMM proposals provide a fragmentary picture of digital maturity in terms of the concept of DT [16–18]. Existing DMMs lack consistency, clarification, and applicability [17–21]. Thus, there is a significant need for an application-oriented DMM that takes into account the holistic and socio-technical nature of DT [13, 16, 18, 21]. For this, it is first crucial to know what an organization should assess internally [3, 5, 13, 22–24]. Through this call for further research, we aim to develop a socio-technical framework that maps a complete picture of DT and helps (re)designing DMMs. With a systematic literature review and a Delphi study, we aim to answer the following research question: *Which set of organizational dimensions and characteristics maps the holistic and socio-technical nature of digital transformations in organizations?*

Our literature-based and empirically-validated framework contributes to meeting the research need by pointing out the full extent of DT-related characteristics within organizations, regardless of their industries or sectors. This serves as a pattern for (re)designing DMMs. The paper is structured as follows. The next section gives a brief overview of relevant research background around DT and digital maturity before we outline the research methodology of our study in the following section. In the subsequent section, we present our preliminary results by answering our research questions. The last section provides an overview of the contributions, implications, and limitations of our work.

## 2    Research Background

DT has become a significant keyword with a variety of definitions and relationships with other similar concepts in the literature. The consensus is that DT is a process of organizational change induced and driven by new digital technologies. It has an organization-wide impact on a multitude of dimensions and their components of an organization, such as business models, operational processes, or customer touchpoints [2, 3, 5, 9, 11, 25, 26]. A DT brings forth a new organizational identity [4]. Therefore, a DT differs in the degree of change to the concepts of digitalization and IT-enabled organizational transformation. Digitalization changes simple business processes and operations, whereas IT-enabled organizational transformation furthermore reinforces the organization's value proposition [3]. A DT goes beyond that by affecting the whole organization [3, 11, 27]. It leverages digital resources to create differentiated value [28]. Induced changes are felt across the socio-technical structures of the organization [5, 7, 8]. Moreover, the changes map the business environment's complexity and the disruptive impact of digital technologies at the individual, organizational and societal levels [5]. Therefore, proper management is needed to guide properly through the DT process [29]. As an organization, it is crucial to embed transformation objectives into the business strategy [30]. Additionally, DT progress needs to be monitored and assessed to be able to take targeted actions [12, 25, 29]. Thus, organizations need to know which organizational characteristics contribute to DT and therefore need to be taken into account [3, 5]. However, organizations find it challenging to assess an internal status quo [25]. In other words, maturity is not or only vaguely determined.

Maturity, in a broader sense, reflects evolutionary progress made in business objectives or capabilities from a start to a desired respectively defined end-stage [15, 24]. In IS research, maturity is viewed as a measure for assessing organizational characteristics [15]. In a DT context, the term digital maturity is used and specifically expresses which DT progress has already been realized [9, 31]. As stated above, a DT influences socio-technical structures. Therefore, digital maturity goes beyond a purely technological perspective, but also comprises a managerial aspect. It describes what progress have already been achieved within different organizational dimensions [9, 18].

To monitor transformation progress, maturity models are useful and well-established tools in IS research [13, 20, 22, 24, 32, 33]. Maturity models are valuable because they outline typical, predictable, or desired paths of potential trajectories of specific organizational characteristics [34]. Hence, maturity models also appeal in the digital context [20]. Such digital maturity models (DMMs) consist of measurable and relevant characteristics that can be assigned to organizational dimensions and grouped if necessary into components [12, 23]. Researchers have since been engaged in developing DMMs, applying or validating existing ones, mostly in a top-down approach, and analyzing them on a meta-level [33]. Developers of DMMs can draw on various proposals for DMM development methods in the literature [23, 34, 35].

Nevertheless, improvements to enhance the quality of existing DMMs in IS research [22, 33] can be sought out because most DMMs lack consistency, clarification, and applicability [17–21]. Regarding accuracy, Becker et al. [34] see an increasing inconsistency and low accuracy when it comes to DMM development and validation approaches. In particular, and in most cases, the procedure of data collection is not described transparently and the measurement validity is deficient [21]. Gökalp and Martinez [19] add that many DMMs were not published in academic peer-reviewed articles. In terms of clarification, existing DMMs convey an incomplete picture of digital maturity because they only address one or a few dimensions [17, 18, 20]. Moreover, DMM dimensions have rarely been conceptualized and specified in detail [18, 19, 22]. Regarding applicability, Gökalp and Martinez [19] conclude that existing DMMs do not show an integrated and empirically validated approach for application.

With consideration of the stated shortcomings of DMMs, a research gap becomes visible. It requires DMMs that meet the holistic and socio-technical nature of a DT by including all measurable and transformation-relevant characteristics of an organization [13, 18, 21]. Such a DMM should be described in such detail that its containing dimensions possess greater depth. In this context, it is proposed to first understand the complexity of DT by systematically identifying characteristics that are relevant for mastering DT. This is where our paper comes in providing a socio-technical framework as a pattern for (re)designing DMMs.

## 3 Research Methodology

We use a conceptual framework for systematizing the research results [36]. Methodically, we orientate on acknowledged maturity model development procedures from previous IS research and on their requirements for the design of a maturity model [23, 34, 35]. In this way, we aim to develop a framework that can be used as a pattern within the procedure of formulating a DMM's architecture and content. When identifying the framework's content, scientific literature proposes to conduct an extensive review of DMM and DT literature. The literature-based findings need to be evolved and tested for "comprehensiveness, consistency, and problem adequacy" [34] by other methods [23]. Expert groups in the context of Delphi studies are one suitable method [23, 34, 35]. Thus, we structure our approach in a literature review and empirical revision phase. In the first phase, we conducted a systematic literature review based on the suggestions of Webster and Watson [37]. In the second phase, we conducted a Delphi study comprising a survey of an expert panel over four rounds.

When reviewing the literature, we identified relevant context- as well as object-related keywords and then set up an adequate search string: *"digital transformation" AND ((maturity OR assessment OR readiness OR capability OR "capability maturity" OR "maturity grid" OR "stage of growth") AND (model OR framework OR map))*. Due to consensual differences in the concepts of DT, IT-enabled organization, digitalization and digitization in literature [3, 4] as well as our research focus on DT, we only included the term *digital transformation*. Previously, the term *digital maturity* was additionally included in the search string. But it was discarded due to the high number of non-fitting literature that mostly deals with the adoption of single technologies and that are out of the context of DT. The search string was then used for a query in the databases *Scopus, EBSCOhost*, as well as *IEEE Xplore*. Here, we only included peer-reviewed journal or conference articles in English that are available online. Duplicates were excluded. In an initial review of titles, keywords, and abstracts of 228 hits, we checked on accessibility and whether these articles focus on DT and digital maturity. After excluding non-relevant articles, we reviewed the full texts of 151 articles, excluded other irrelevant articles, and could ultimately extract organizational characteristics related to transformation processes, especially to a DT, from 39 articles. To assign the identified characteristics to organizational dimensions, we reviewed suitable frameworks encompassing dimensions in the context of DT and change management. We see common ground best represented by the 7S framework (7S) enabling a holistic and socio-technical view of essential organizational dimensions. The 7S was formulated and tested by Waterman et al. [38] and initially intended for McKinsey's business consulting purposes. Since then, the framework was used and adapted in various areas of research, even in the context of maturity assessment and digitalization [39–44]. Waterman et al. [38] claim that organizational change is only effective and successful if all relevant dimensions of an organization are considered. These 7S dimensions can be summarized as follows: *Strategy* describes all strategic actions in interdependence with external circumstances, which are essential for the business ability of the organization. *Structure* is the visible and formalized skeleton in the form of departments, teams, and tasks of an organization. *Systems* of an organization consist of the technical infrastructure as well as formal and informal processes that support other activities within the other dimensions. *Staff* includes all activities that impact the *Structure* and culture of the workforce and that shape the employee skill set. *Style* of an organization is affected by the leadership style and all related activities and behavior of managers and employees. *Skills* are understood as all skills, competencies, and knowledge that exist on an individual, team, and organizational level. *Superordinate Goals* shape the corporate culture and act as a guideline for daily work. [38] When having the complete list of organizational characteristics after reviewing the articles' full-texts, we derive concepts according to Webster and Watson [37]. We aggregated duplicate or similar characteristics and finally assigned them to an appropriate dimension of the 7S. The decision process of assigning took place in workshops among the authors.

Our resulting theoretical framework was then empirically validated and refined using the Delphi method in the second phase. To do so, anonymous expert opinions were obtained in several rounds. In general, the Delphi method can be used for answering complex questions and for elaborating future directions [45]. It is an iterative approach in which multiple surveys are conducted until a consensus among experts is reached [45]. Delphi studies are well established in IS research [46] and have been successfully used in maturity model research [47, 48]. At the beginning of a Delphi study, a panel of experts is established, who can provide information about the topic area [45]. According to the literature, the expert panel is appropriate with a number of 10–18 participants, and the experts must remain anonymous among themselves. This is to prevent conflicts within the group as well as peer pressure [46]. The experts then evaluate the given topics in several rounds. After each round, the results of all participants are consolidated. On this basis, iterative adaptions are made, which are finally approved by the experts. A broad sample increases the chance of capturing different impressions in the data [49]. Therefore, we selected twelve experts with different fields of expertise, backgrounds, ages, and professional experience for our Delphi study. Participants worked either in academia (3) or practice (9) in Germany. Those working in academia qualified for our survey by conducting their research in the field of DT and/or having already developed maturity models. The participants working in companies have either practical experience in the area of DT, in the realization and development of digitization projects, in developing maturity models, or are working in an IT division. Our sample contains a broad spectrum of working experiences and ages of the participants. Professional experience varies from five or fewer years up to 16–20 years. A similar pattern can be noted regarding the age of the participants, which is between 21 and 50 years. In the four rounds of our Delphi study, 9–12 experts participated in each round. After the experts were determined, the study was conducted in four rounds. Each round consisted of a questionnaire provided through an online survey tool. The findings of the previous literature review served as the basis for the surveys in the form of aggregated dimensions and characteristics. Each round started with relevant background information. Rounds 1 and 2 focused on the dimensions, while rounds 3 and 4 dealt with the characteristics of the respective dimensions. **Round 1** began with definitions of DT and DMM to achieve a common understanding among all experts. The experts were asked to evaluate the seven dimensions, which were defined based on the literature review, and summarize all relevant aspects of an organization in a very abstract way. Each dimension was presented using a definition. For each dimension, the experts could choose to *Retain* (the dimension should be kept exactly as it is presented), *Adapt* (the dimension should be changed or extended), or *Drop* (the dimension should be completely removed from the framework). For this purpose, a selection box was provided for the experts to click on. In a separate field, the experts were allowed to make additional comments, which they should use if they had adaptions and/or additional requests. In addition, the experts were asked whether the bundle of dimensions were complete or whether dimensions should be added. In preparation for the next round, necessary adaptions were made based on the results from the first round. A dimension is considered to be confirmed if it has a retention rate of at least a two-thirds majority. For a dimension to be completely dropped, a drop rate of at least a two-thirds majority is required. If these rates are just not met, it will be decided on a case-by-case basis which changes are necessary according to the experts' change recommendations. The adapted dimensions were presented to the experts for validation in **round 2**. The experts again had the three selection options *Retain*, *Adapt* and *Drop* as well as a comment field available. Additionally, the experts were asked to weigh the dimensions according to relevance. A maximum of 100 points was available, which had to be assigned to the individual dimensions by the experts. The sum of all the points awarded had to total 100. In **round 3**, the focus was on the characteristics of the dimensions that were assessed by experts. Within each dimension, a decision had to be made for each characteristic, whether it should be retained, adapted, or dropped. Again, the selection fields, including a comment field, were available. The results from round 3 were evaluated analogously to the first rounds, and the necessary adaptations of the characteristics were made. Finally, in **round 4**, the adapted characteristics were presented to and evaluated by the expert panel. After this round, the Delphi study could be concluded as a consensus was reached among the experts. Thus, the last minor adaptions were made, resulting in the finalization of the dimensions and characteristics.

## 4   Results

Due to the conducted systematic literature review in our first phase, we initially could extract a sum of 698 DT-related organizational characteristics from 39 relevant articles. These could be completely assigned to at least one dimension of the 7S. All articles addressed at least one 7S dimension. However, it is noticeable that most articles (35) have the dimension *Systems* under consideration, followed by *Strategy* (29) and *Style* (26). Three articles cover all seven dimensions, two of which aim at a state-of-the-art and one on a literature-based development of a DMM. Regarding the articles' research focus, it is striking that most of the literature (28) is to develop a DMM, whereas only a few DMMs are applied (5) or validated (4). In addition, articles with a focus to elaborate a state-of-the-art (3) were identified. Of the 698 characteristics, we aggregated duplicates and similar ones into a total of 48 different characteristics. This results in the design of a literature-based framework that we empirically revised in our second phase. The holistic approach of our framework was positively welcomed by the expert panel of our Delphi study. According to the expert panel, it covers the main facets of DT to map its complexity and impacts on practice within organizations. Our framework provides a sufficient overview of relevant aspects, which in turn leads to an optimized decision-making process. Nevertheless, "the model covers a lot, but there can be added more facets, which might be partly subordinated to other dimensions" [Delphi study]. The expert panel points out that it was difficult to delineate individual dimensions and characteristics because of their close relationships that are expressed intentionally by the 7S's definitions. Due to losses in terms of understanding of the dimensions and clarity about the completeness, the literature-based framework was subjected to revision. The empirically-revised framework is introduced in *Table* 1 and substantial revisions are described below.

**Table 1.** Socio-technical framework for DT (literature-based and empirically-revised).

| Dimension and characteristics | Literature basis |
|---|---|
| **Strategy:** Existence of clearly defined digital vision I Organization synchronized with digital vision I Existence of systematic developed digital strategy I Holistic execution of digital strategy I Development and offering of digital services or products to strengthen the existing business model or to enable new business models I Internal integration of digital technologies I Budgeting for digital innovation considering potential qualitative and quantitative benefits I Focus on customer value; Involvement of customers in the innovation process I Cooperation with business partners in a digital ecosystem I Conducting technological trend analysis I Corporate governance providing standards, ensuring ethics and compliance with laws | [3, 18, 50–76] |
| **Structure & Process:** Tasks, responsibilities, and competencies defined to support staff in execution I Coordination of centralized and decentralized digitalization efforts I Adequate resource allocation for digitization and transformation activities I Decentralized approach resulting in an extensive scope of action for divisions, departments, and working groups I Agile and flexible organizational configuration I Exchange across departments, business units, and organization borders I Collaboration in multidisciplinary teams I Data-driven product and service development in line with strategy I Data-driven resource planning processes I Digitally-modeled operations resulting in higher process transparency I Automated operational processes | [3, 18, 50–63, 65–83] |
| **Technology & Data:** Harmonized and resilient technology landscape I Communicating and interoperable equipment and installations I Interoperable and data-driven mobile devices & embedded systems I Automated and customizable application and service systems I Automated and expedient data acquisition and storage I Ensuring high data quality I Intra- and inter-organizational data integration and sharing I Ensured data security, protection, and ownership I Data analysis for operational and strategic purposes | [18, 53, 55–59, 61, 62, 64, 65, 67, 68, 70, 72, 75, 77–80, 82, 83] |

(*continued*)

**Table 1.** (*continued*)

| Dimension and characteristics | Literature basis |
| --- | --- |
| **Culture:** Innovative ideas are contributed by the entire staff I The risk of failure is taken to realize new ideas I Willingness to learn from errors I Organizational knowledge is shared and preserved internally I Openness to intra- and inter-organizational collaborations I Existence of transformational leadership style I Information is shared within the organization I Participative interaction between staff I Awareness of new digital developments I Openness to new technologies | [18, 52, 54–60, 62, 64, 65, 68, 69, 71, 72, 75, 79–81, 83–86] |
| **Skills:** Ability for teamwork I Ability to learn continuously I Ability to use digital tools I Technical knowledge is available I Data are handled I Staff is hired based on needed expertise I Analyzes show whether the necessary skills are sufficient I Staff is trained | [18, 51, 52, 54, 56–59, 62, 64, 68, 69, 72, 74–76, 80, 82, 84, 86] |

The first dimension *Strategy* comprises all activities that align the vision, the guidelines, and the business model with the political-legal, economic, social, and ecological circumstances of the business environment. In the origin 7S, a distinction is made between *Strategy* and *Superordinate Goals*. Many practitioners and even some scientists in our Delphi study had difficulties in differentiating the *Superordinate Goals* from the concept of *Strategy* and *Culture* and therefore pleaded for a merger. We have followed these suggestions as we pursue the goal of an application-oriented and user-friendly framework and separated the *Superordinate Goals* into *Strategy* and *Culture*. *Superordinate Goals* shape on the one hand the corporate culture and act as a guideline for daily work, and on the other hand, they give "notions of future direction" [38]. These notions are particularly shaped by aspects like the vision and corporate governance, which in turn is strongly linked to the strategy as many participants noted.

The second dimension called *Structure & Processes* comprises all visible and measurable components and activities concerning the organization's internal processes as well as its structure, such as hierarchies, departments, and teams. This dimension is a merger of *Structure* and a component of *Systems* belonging to the initial framework based on the 7S [38]. The *Structure* is the visible and formalized skeleton of an organization. Employees are aggregated into departments and tasks are divided into subtasks [38]. *Systems* includes partially formal and informal processes that support other activities within the other dimensions [38]. The original term *Systems* was misleading and required a separation of the technological and process component. According to the expert panel, a process component was not sufficiently clear, although especially value creation processes play an essential role in ensuring organizational alignment at a strategic and operational level. Because they saw a close relationship between internal processes and structure, these were therefore merged.

The dimension *Technology & Data* has been reformulated by the stated separation of *Systems* as well as by the emphasis of the expert panel on the importance of data management in the context of DT. Thus, the dimension comprises the composition, inter-action, and functionalities of technical resources required for processing information and data, as well as all activities within the scope of data management. Technical resources are hardware, software, and communication networks whose aggregation represents the technology landscape of an organization.

*Culture* comprises the overall values, norms, mindsets of the workforce, and leader-ship styles of management that determine behavior at an individual, team, and organiza-tional level. This dimension was called *Style* in the underlying 7S. However, the expert panel disagreed with this designation and required a renaming. The adapted term was confirmed in the second round with a high retention rate. The characteristics of the initial dimension *Style* were simply transferred to *Culture* in the first step. The characteristics of the initial dimension were all confirmed by the experts. However, characteristics had to be added based on the results of the Delphi study.

The dimension *Skills* comprises the availability, preservation, and development of knowledge, competencies, and experience at an individual, team, and organizational level. In the underlying 7S, a differentiation is made between *Skills* and *Staff*. However, the expert panel identified strong overlaps between the two dimensions, which should not be considered separately at all. Within *Skills*, "[…] strong overlaps with the item *Staff*" [Delphi study] were seen, while within *Staff* the experts also found it "not quite clearly distinguishable from *Skills*" [Delphi study]. Thus, the two dimensions were combined overarchingly into *Skills*. The adapted dimension was confirmed by the experts in the round 2 so that the characteristics of the dimension were focused on in the third and fourth rounds. The characteristics of the initial dimensions were first combined and adaptions were made after the experts' evaluations in round 3.

In sum, two dimensions were merged, one dimension was split, and assigned to two other dimensions respectively in rounds 1 and 2 of our Delphi study. Concerning the weighting of the dimensions, the expert panel achieved an average of almost equal weighting. *Strategy* took rounded 22%, *Structure & Processes* 15%, *Technology & Data* 22%, *Culture* 22%, and *Skills* 19%. During rounds 3 and 4, out of a total of 48 character-istics, 18 were retained, 26 were adapted, four were dropped and six were supplemented. Thus, we obtained an empirically-revised framework consisting of five dimensions to which we link a total of 50 characteristics.

## 5 Discussion and Conclusion

The starting point of our research was the need for an application-oriented DMM that takes into account the holistic and socio-technical nature of DT [13, 16, 18, 21]. Thus, we aimed to provide a framework that improves DT management by serving as a pattern for (re)designing DMMs. Existing DMMs convey an incomplete picture of DT's complexity by only addressing a few dimensions [17, 18, 20]. This limits generalizability, which is why we ensure holism by viewing organizations as socio-technical systems [6], so we take a perspective on social and technical implications [87].

We chose a two-phase methodological approach consisting of a literature review and a Delphi study. First, we built a preliminary theoretical framework based on important findings from previous IS research. Existing DMMs and further research on DT and digital maturity provided a suitable basis for this purpose [33, 34]. To classify DT-related characteristics extracted from the literature, we used a holistic organizational model. The 7S [38] is an appropriate basis that helps in mapping a complex subject area, such as DT. In addition, related research indicates that the 7S can also be adapted and used in distinct actual topics in the context of maturity assessment and digitalization [39–44, 88]. By using the 7S as a pattern in the literature review, the stated research need was confirmed. We identified only one developed DMM that covers all dimensions. Still, this model is at a theoretical level and has not yet been empirically validated. In terms of validity and the required application orientation [19], we went one step further and had our interim results tested and refined by an expert panel in a Delphi study. The anonymous nature of the Delphi study provided creative input and enriched our theoretical framework [46]. The method showed us that formulated dimensions and characteristics were able to be refined. The 7S was compressed into five dimensions and *Technology & Data* were given a stronger role by name than before. This might be due to the 7S's age as well as the driving character of technologies for a DT [2–5]. Moreover, characteristics that were not depicted in the literature were added by the expert panel. With our two-phase methodology accepted in IS research [48], we achieved to develop a socio-technical framework. While DDMs in IS research do not yet map the holistic and socio-technical nature of a DT, our framework provides a comprehensive composition of technical and social dimensions and characteristics. Thus, a socio-technical perspective is taken and the organization is understood as a socio-technical system [6]. This perspective considers both technical and social implications of DT and means that emerging changes are a result of the interactions between the two [87]. By using a two-layered framework architecture, we keep the recommended architecture for maturity models [23] to ensure the balance of representing the complexity of DT, on the one hand, and simplicity for reuse, on the other hand. In addition, adherence to proven development procedure models [23, 34, 35] supports further research. The containing dimensions and characteristics considered in interaction and dependence, our framework reflects the necessary socio-technical view on a DT [5, 7, 8]. The expert panel's weighting supports this perspective with its appraisal that equal attention should be given to all dimensions for a DT.

Nevertheless, our research has limitations. It remains to be emphasized that our socio-technical framework is not a complete DMM. The review of the small number of identified DT- and DMM-related articles only gives an excerpt of all developed DMMs in the literature. Extending the search field by replacing the term *digital transformation* by *digit\** to include all other concepts in the digital context could provide more comprehensive results. Nevertheless, this would avert the necessary research focus on DT. Moreover, our design decision process of assigning and aggregating extracted characteristics might have a subjective character due to the common professional background of the authors. Involving an interdisciplinary and larger group in this process could lead to more objective results here, as well. We aimed to dissolve these limitations by empirically testing our preliminary and theoretical results with a 12-member expert panel from academia and practice. This approach is also beset with limitations that are rooted in the nature of Delphi studies. The framework's refinements and coding procedure are based on the perceptions of a group of experts which reduces representativeness [89].

Still, our framework helps facilitate the normalization of DT [13] and accelerate DT in practice by building an understanding and awareness of DT managers for key characteristics in their organization when dealing integrally with DT. Our results guide where DT can progress and suitable actions can be taken in organizations. Further, there is no universal answer to the research questions for an IS problem. Thus, our socio-technical framework reflects one contribution to the ongoing research and is a starting point for a community-wide discussion. A refinement of the framework within for example an iterative maturity model development phase [34] or, more systematically, within a design science research cycle [90] is also feasible. We invite researchers to evaluate, adapt, or extend our framework in further research. In particular, our socio-technical framework should be tested in the context of a DMM design. At this, a methodological basis can also be design science research as previous work around maturity models have done [91, 92]. In addition, reference models such as the Capability Maturity Model Integration that has proven itself in IS research [82] can serve as a suitable basis for integrating our framework. In this way, the holistic and socio-technical nature of DT will be represented in maturity models, so digital transformation maturity models will be supplemented.

# References

1. Osmundsen, K., Iden, J., Bygstad, B.: Digital transformation: drivers, success factors, and implications,16 (2018)
2. Hanelt, A., Bohnsack, R., Marz, D., Antunes Marante, C.: A systematic review of the literature on digital transformation: insights and implications for strategy and organizational change. J. Manag. Stud. **58**, 1159–1197 (2021)
3. Verhoef, P.C., et al.: Digital transformation: a multidisciplinary reflection and research agenda. J. Bus. Res. **122**, 889–901 (2021)
4. Wessel, L., Baiyere, A., Ologeanu-Taddei, R., Cha, J., Blegind Jensen, T.: Unpacking the difference between digital transformation and IT-enabled organizational transformation. JAIS **22**, 102–129 (2021)
5. Vial, G.: Understanding digital transformation: a review and a research agenda. J. Strateg. Inf. Syst. **28**, 118–144 (2019)
6. Leonardi, P.M.: Materiality, sociomateriality, and socio-technical systems: what do these terms mean? How are they related? Do we need them? SSRN J., 25–48 (2012)
7. Hess, T., Matt, C., Benlian, A., Wiesböck, F.: Options for formulating a digital transformation strategy. MIS Q. Exec. **15**, 123–139 (2016)
8. Yoo, Y., Lyytinen, K.J., Boland, R.J., Berente, N.: The next wave of digital innovation: opportunities and challenges: a report on the research workshop "digital challenges in innovation research." SSRN J. (2010)
9. Chanias, S., Hess, T.: How digital are we? Maturity models for the assessment of a company's status in the digital transformation. LMU München (2016)
10. Dregger, J., Niehaus, J., Ittermann, P., Hirsch-Kreinsen, H., ten Hompel, M.: The digitization of manufacturing and its societal challenges: a framework for the future of industrial labor. In: 2016 IEEE International Symposium on Ethics in Engineering, Science and Technology (ETHICS), Vancouver, BC, Canada, pp. 1–3. IEEE (2016)
11. Kane, G.C., Palmer, D., Phillips, A.N., Kiron, D., Buckley, N.: Strategy, not technology, drives digital transformation. MIT Sloan Management Review and Deloitte University Press, 14 July 2015

12. Berghaus, S.: Stages in digital business transformation: results of an empirical maturity study, 18 (2016)
13. Carroll, N.: Theorizing on the normalization of digital transformations. In: Proceedings of the 28th European Conference on Information Systems (ECIS) (2020)
14. Klötzer, C., Pflaum, A.: Toward the development of a maturity model for digitalization with-in the manufacturing industry's supply chain. In: Proceedings of the 50th Hawaii International Conference on System Sciences, pp. 4210–4219 (2017)
15. Rosemann, M., de Bruin, T.: Towards a business process management maturity model. In: ECIS 2005 Proceedings of the Thirteenth European Conference on Information Systems, pp. 1–12. Verlag and the London School of Economics (2005)
16. Aguiar, T., Gomes, S.B., da Cunha, P.R., da Silva, M.M.: Digital transformation capability maturity model framework. In: 2019 IEEE 23rd International Enterprise Distributed Object Computing Conference (EDOC), Paris, France, pp. 51–57. IEEE (2019)
17. Schwer, K., Hitz, C., Wyss, R., Wirz, D., Minonne, C.: Digital maturity variables and their impact on the enterprise architecture layers. Probl. Perspect. Manag. **16**, 141–154 (2018)
18. Teichert, R.: Digital transformation maturity: a systematic review of literature. Acta Univ. Agric. Silvic. Mendelianae Brun. **67**, 1673–1687 (2019)
19. Gökalp, E., Martinez, V.: Digital transformation capability maturity model enabling the assessment of industrial manufacturers. Comput. Ind. **132**, 103522 (2021)
20. Schallmo, D.R.A., Lang, K., Hasler, D., Ehmig-Klassen, K., Williams, C.A.: An approach for a digital maturity model for SMEs based on their requirements. In: Schallmo, D.R.A., Tidd, J. (eds.) Digitalization. MP, pp. 87–101. Springer, Cham (2021). https://doi.org/10.1007/978-3-030-69380-0_6
21. Thordsen, T., Murawski, M., Bick, M.: How to measure digitalization? A critical evaluation of digital maturity models. In: Hattingh, M., Matthee, M., Smuts, H., Pappas, I., Dwivedi, Y.K., Mäntymäki, M. (eds.) I3E 2020. LNCS, vol. 12066, pp. 358–369. Springer, Cham (2020). https://doi.org/10.1007/978-3-030-44999-5_30
22. Becker, J., Niehaves, B., Poeppelbuss, J., Simons, A.: Maturity models in IS research. In: 18th European Conference on Information Systems, ECIS 2010 (2010)
23. de Bruin, T., Rosemann, M., Freeze, R., Kulkarni, U.: Understanding the main phases of developing a maturity assessment model. In: ACIS 2005 Proceedings of the American Conference on Information Systems, p. 11 (2005)
24. Lahrmann, G., Marx, F., Mettler, T., Winter, R., Wortmann, F.: Inductive design of maturity models: applying the Rasch algorithm for design science research. In: Jain, H., Sinha, A.P., Vitharana, P. (eds.) DESRIST 2011. LNCS, vol. 6629, pp. 176–191. Springer, Heidelberg (2011). https://doi.org/10.1007/978-3-642-20633-7_13
25. Hansen, A.M., Kraemmergaard, P., Mathiassen, L.: Rapid adaption in digital transformation: a participatory process for engaging IS and business leaders. MIS Q. Exec. **10**, 175–185 (2011)
26. Reis, J., Amorim, M., Melão, N., Matos, P.: Digital transformation: a literature review and guidelines for future research. In: Rocha, Á., Adeli, H., Reis, L.P., Costanzo, S. (eds.) World-CIST'18 2018. AISC, vol. 745, pp. 411–421. Springer, Cham (2018). https://doi.org/10.1007/978-3-319-77703-0_41
27. Amit, R., Zott, C.: Value creation in E-business. Strat. Manag. J. **22**, 493–520 (2001)
28. Bharadwaj, A., El Sawy, O., Pavlou, P., Venkatraman, N.: Digital business strategy: to-ward a next generation of insights. Manag. Inf. Syst. Q. **37**, 471–482 (2013)
29. Pabst von Ohain, B.: Leader attributes for successful digital transformation. In: ICIS 2019 Proceedings (2019)
30. Matt, C., Hess, T., Benlian, A.: Digital transformation strategies. Bus. Inf. Syst. Eng. **57**, 339–343 (2015). https://doi.org/10.1007/s12599-015-0401-5

31. Kane, G.C., Palmer, D., Phillips, A.N., Kiron, D., Buckley, N.: Achieving digital maturity. MIT Sloan Management Review and Deloitte University Press (2017)

32. Mettler, T., Rohner, P.: Situational maturity models as instrumental artifacts for organizational design. In: Proceedings of the 4th International Conference on Design Science Research in Information Systems and Technology - DESRIST 2009, Philadelphia, Pennsylvania, p. 1. ACM Press (2009)

33. Wendler, R.: The maturity of maturity model research: a systematic mapping study. Inf. Softw. Technol. **54**, 1317–1339 (2012)

34. Becker, J., Knackstedt, R., Pöppelbuß, J.: Developing maturity models for IT management: a procedure model and its application. Bus. Inf. Syst. Eng. **1**, 213–222 (2009). https://doi.org/10.1007/s12599-009-0044-5

35. van Steenbergen, M., Bos, R., Brinkkemper, S., van de Weerd, I., Bekkers, W.: The design of focus area maturity models. In: Winter, R., Zhao, J.L., Aier, S. (eds.) DESRIST 2010. LNCS, vol. 6105, pp. 317–332. Springer, Heidelberg (2010). https://doi.org/10.1007/978-3-642-13335-0_22

36. Leshem, S., Trafford, V.: Overlooking the conceptual framework. Innov. Educ. Teach. Int. **44**, 93–105 (2007)

37. Webster, J., Watson, R.T.: Analyzing the past to prepare for the future: writing a literature review. MIS Q. **26**, xiii–xxiii (2002)

38. Waterman, R.H., Peters, T.J., Phillips, J.R.: Structure is not organization. Bus. Horiz. **23**(3), 14–26 (1980)

39. Kocaoglu, B., Demir, E.: Maturity assessment in the technology business within the Mckinsey's 7S framework. Res. J. Bus. Manag. **6**, 158–166 (2019)

40. Versteck, M.: The Seven S Framework and its use as an assessment tool. Innov. High. Educ. **13**, 106–116 (1989). https://doi.org/10.1007/BF00889744

41. Zapukhliak, I., Zaiachuk, Y., Polyanska, A., Kinash, I.: Applying fuzzy logic to assessment of enterprise readiness for changes. Manag. Sci. Lett. **9**(13), 2277–2290 (2019). https://doi.org/10.5267/j.msl.2019.7.026

42. Hanafizadeh, P., Ravasan, A.Z.: A McKinsey 7S model-based framework for ERP readiness assessment. Int. J. Enterp. Inf. Syst. **7**, 23–63 (2011)

43. Kaplan, R.S.: How the balanced scorecard complements the McKinsey 7-S model. Strategy Leadersh. **33**, 41–46 (2005)

44. Krikhaar, R., Mermans, M.: Software development improvement with SFIM. In: Münch, J., Abrahamsson, P. (eds.) PROFES 2007. LNCS, vol. 4589, pp. 65–80. Springer, Heidelberg (2007). https://doi.org/10.1007/978-3-540-73460-4_9

45. Loo, R.: The Delphi method: a powerful tool for strategic management. Policing Int. J. Police Strat. Manag. **25**, 762–769 (2002)

46. Okoli, C., Pawlowski, S.D.: The Delphi method as a research tool: an example, design considerations and applications. Inf. Manag. **42**, 15–29 (2004)

47. Serral, E., Stede, C.V., Hasic, F.: Leveraging IoT in retail industry: a maturity model. In: 2020 IEEE 22nd Conference on Business Informatics (CBI), Antwerp, Belgium, pp. 114–123. IEEE (2020)

48. Stoiber, C., Schönig, S.: Digital transformation and improvement of business processes with Internet of Things: a maturity model for assessing readiness. Presented at the Hawaii International Conference on System Sciences (2022)

49. Yin, R.K.: Case Study Research: Design and Methods. SAGE, Los Angeles and London and New Delhi and Singapore and Washington, DC (2014)

50. Exner, K., Balder, J., Stark, R.: A PSS maturity self-assessment tool. Procedia CIRP **73**, 86–90 (2018)

51. Santos, R.C., Martinho, J.L.: An Industry 4.0 maturity model proposal. JMTM **31**, 1023–1043 (2019)

52. Colli, M., Madsen, O., Berger, U., Møller, C., Wæhrens, B.V., Bockholt, M.: Contextualizing the outcome of a maturity assessment for Industry 4.0. IFAC-PapersOnLine **51**, 1347–1352 (2018)

53. Isaev, E., Korovkina, N., Tabakova, M.: Evaluation of the readiness of a company's IT department for digital business transformation. Bus. Inform. **2018**, 55–64 (2018)

54. Mittal, S., Romero, D., Wuest, T.: Towards a smart manufacturing maturity model for SMEs (SM3E). In: Moon, I., Lee, G.M., Park, J., Kiritsis, D., von Cieminski, G. (eds.) APMS 2018. IAICT, vol. 536, pp. 155–163. Springer, Cham (2018). https://doi.org/10.1007/978-3-319-99707-0_20

55. Zaoui, F.: A triaxial model for the digital maturity diagnosis. IJATCSE **9**, 433–439 (2020)

56. Sahu, N., Deng, H., Molla, A.: A capability based framework for customer experience focused digital trans-formation. In: School of Business Information Technology and Logistics RMIT University, AU (ed.) Australasian Conference on Information Systems 2018. University of Technology, Sydney (2018)

57. Al-Sai, Z.A., Abdullah, R., Husin, M.H.: Critical success factors for big data: a systematic literature review. IEEE Access. **8**, 118940–118956 (2020)

58. Gamache, S., Abdul-Nour, G., Baril, C.: Development of a digital performance assessment model for Quebec manufacturing SMEs. Procedia Manufact. **38**, 1085–1094 (2019)

59. Gurbaxani, V., Dunkle, D.: Gearing up for successful digital transformation. MISQE **18**, 209–220 (2019)

60. Seitz, J., Burosch, A.: Digital value creation. In: 2018 IEEE International Conference on Engineering, Technology and Innovation (ICE/ITMC), Stuttgart, pp. 1–5. IEEE (2018)

61. Pfenning, P., Eibinger, H.C., Rohleder, C., Eigner, M.: A comprehensive maturity model for assessing the product lifecycle. In: Nyffenegger, F., Ríos, J., Rivest, L., Bouras, A. (eds.) PLM 2020. IAICT, vol. 594, pp. 514–526. Springer, Cham (2020). https://doi.org/10.1007/978-3-030-62807-9_41

62. Sanchez, M.A.: A framework to assess organizational readiness for the digital transformation. Dimensión Empresarial **15**, 27–40 (2017)

63. Muñoz, L., Avila, O.: A model to assess customer alignment through customer experience concepts. In: Abramowicz, W., Corchuelo, R. (eds.) BIS 2019. LNBIP, vol. 373, pp. 339–351. Springer, Cham (2019). https://doi.org/10.1007/978-3-030-36691-9_29

64. Soni, F.S.G., Nugroho, H.: An evaluation of E-readiness cloud computing service model adoption on Indonesian higher education. In: 2018 6th International Conference on Information and Communication Technology (ICoICT), pp. 28–33 (2018)

65. Vuksanović Herceg, I., Kuč, V., Mijušković, V.M., Herceg, T.: Challenges and driving forces for Industry 4.0 implementation. Sustainability **12**, 4208 (2020)

66. Muncinelli, G., de Lima, E.P., Deschamps, F., da Costa, S.E.G., Cestari, J.M.A.P.: Components of the preliminary conceptual model for process capability in LGPD (Brazilian Data Protection Regulation) context. In: Pokojski, J., Gil, M., Newnes, L., Stjepandić, J., Wognum, N. (eds.) Advances in Transdisciplinary Engineering. IOS Press (2020)

67. Buhulaiga, E.A., Telukdarie, A., Ramsangar, S.J.: Delivering on Industry 4.0 in a multinational petrochemical company: design and execution. In: 2019 International Conference on Fourth Industrial Revolution (ICFIR), Manama, Bahrain, pp. 1–6. IEEE (2019)

68. Ifenthaler, D., Egloffstein, M.: Development and implementation of a maturity model of digital transformation. TechTrends **64**(2), 302–309 (2019). https://doi.org/10.1007/s11528-019-00457-4

69. Sousa-Zomer, T.T., Neely, A., Martinez, V.: Digital transforming capability and performance: a microfoundational perspective. IJOPM **40**, 1095–1128 (2020)

70. Schmitt, P., Schmitt, J., Engelmann, B.: Evaluation of proceedings for SMEs to conduct I4.0 projects. Procedia CIRP **86**, 257–263 (2019)

71. Salinas-Navarro, D.E., Garay-Rondero, C.L.: Experiential learning in industrial engineering education for digital transformation. In: 2019 IEEE International Conference on Engineering, Technology and Education (TALE), Yogyakarta, Indonesia, pp. 1–9. IEEE (2019)
72. González-Varona, J.M., Acebes, F., Poza, D., López-Paredes, A.: Fostering digital growth in SMEs: organizational competence for digital transformation. In: Camarinha-Matos, L.M., Afsarmanesh, H., Ortiz, A. (eds.) PRO-VE 2020. IAICT, vol. 598, pp. 237–248. Springer, Cham (2020). https://doi.org/10.1007/978-3-030-62412-5_20
73. Moura, L.R., Kohl, H.: Maturity assessment in Industry 4.0 – a comparative analysis of Brazilian and German companies. Emerg. Sci. J. **4**, 365–375 (2020)
74. Ramantoko, G., Faitmah, L.V., Pratiwi, S.C., Kinasih, K.: Measuring digital capability maturity: case of small-medium Kampong-digital companies in Bandung. J. Soc. Sci. Humanit. **26**, 215–230 (2018)
75. Schumacher, A., Nemeth, T., Sihn, W.: Roadmapping towards industrial digitalization based on an Industry 4.0 maturity model for manufacturing enterprises. Procedia CIRP **79**, 409–414 (2019)
76. Hamidi, S.R., Aziz, A.A., Shuhidan, S.M., Aziz, A.A., Mokhsin, M.: SMEs maturity model assessment of IR4.0 digital transformation. In: Lokman, A., Yamanaka, T., Lévy, P., Chen, K., Koyama, S. (eds.) KEER 2018. AISC, vol. 739, pp. 721–732. Springer, Singapore (2018). https://doi.org/10.1007/978-981-10-8612-0_75
77. Gökalp, E., Şener, U., Eren, P.E.: Development of an assessment model for Industry 4.0: Industry 4.0-MM. In: Mas, A., Mesquida, A., O'Connor, R.V., Rout, T., Dorling, A. (eds.) SPICE 2017. CCIS, vol. 770, pp. 128–142. Springer, Cham (2017). https://doi.org/10.1007/978-3-319-67383-7_10
78. Leyh, C., Schäffer, T., Bley, K., Forstenhäusler, S.: Assessing the IT and software landscapes of Industry 4.0-enterprises: the maturity model SIMMI 4.0. In: Ziemba, E. (ed.) AITM/ISM -2016. LNBIP, vol. 277, pp. 103–119. Springer, Cham (2017). https://doi.org/10.1007/978-3-319-53076-5_6
79. Setiyawan, J., Gunawan, F., Raharjo, T., Hardian, B.: Application of scrum maturity model: a case study in a telecommunication company. J. Phys. Conf. Ser. **1566**, 012050 (2020)
80. Thornley, C., Carcary, M., Connolly, N., O'Duffy, M., Pierce, J.: Developing a maturity model for knowledge management (KM) in the digital age. In: 16th European Conference on Knowledge Management. University of Ulster, Northern Ireland (2016)
81. Lederer, M., Betz, S., Schmidt, W.: Digital transformation, smart factories, and virtual design: contributions of subject orientation. In: Proceedings of the 10th International Conference on Subject-Oriented Business Process Management - S-BPM One 2018, Linz, Austria, pp. 1–10. ACM Press (2018)
82. De Carolis, A., Macchi, M., Negri, E., Terzi, S.: A maturity model for assessing the digital readiness of manufacturing companies. In: Lödding, H., Riedel, R., Thoben, K.-D., von Cieminski, G., Kiritsis, D. (eds.) APMS 2017. IAICT, vol. 513, pp. 13–20. Springer, Cham (2017). https://doi.org/10.1007/978-3-319-66923-6_2
83. Stich, V., Gudergan, G., Zeller, V.: Need and solution to transform the manufacturing industry in the age of Industry 4.0 – a capability maturity index approach. In: Camarinha-Matos, L.M., Afsarmanesh, H., Rezgui, Y. (eds.) PRO-VE 2018. IAICT, vol. 534, pp. 33–42. Springer, Cham (2018). https://doi.org/10.1007/978-3-319-99127-6_3
84. Nemeth, T., Ansari, F., Sihn, W.: A maturity assessment procedure model for realizing knowledge-based maintenance strategies in smart manufacturing enterprises. Procedia Manufact. **39**, 645–654 (2019)
85. Alos-Simo, L., Verdu-Jover, A.J., Gomez-Gras, J.-M.: How transformational leadership facilitates e-business adoption. IMDS **117**, 382–397 (2017)

86. Schuh, G., Frank, J.: Maturity-based design of corporate culture in the context of Industrie 4.0. In: 2020 International Conference on Technology and Entrepreneurship - Virtual (ICTE-V), San Jose, CA, USA, pp. 1–8. IEEE (2020)

87. Sarker, S., Chatterjee, S., Xiao, X., Elbanna, A.: The sociotechnical axis of cohesion for the IS discipline: its historical legacy and its continued relevance. MISQ **43**, 695–719 (2019)

88. Teh, D., Khan, T., Corbitt, B., Ong, C.E.: Sustainability strategy and blockchain-enabled life cycle assessment: a focus on materials industry. Environ. Syst. Decis. **40**(4), 605–622 (2020). https://doi.org/10.1007/s10669-020-09761-4

89. Ono, R., Wedemeyer, D.J.: Assessing the validity of the Delphi technique. Futures **26**, 289–304 (1994)

90. Hevner, A.R., March, S.T., Park, J., Ram, S.: Design science in information systems research. MIS Q. **28**, 75–105 (2004)

91. Kammerlohr, V., Paradice, D., Uckelmann, D.: A maturity model for the effective digital transformation of laboratories. JMTM (2022)

92. Colli, M., Berger, U., Bockholt, M., Madsen, O., Møller, C., Wæhrens, B.V.: A maturity assessment approach for conceiving context-specific roadmaps in the Industry 4.0 era. Ann. Rev. Control **48**, 165–177 (2019)

# User Journey Map as a Method to Extrapolate User Experience Knowledge from User Generated Reviews

Alberts Pumpurs[(✉)]

Department of Management Information Technology, Riga Technical University, Riga, Latvia
alberts.pumpurs@rtu.lv

**Abstract.** User-generated product reviews are valuable information resources about what users like about the product, their pain points, and overall product use cases. This information is valuable for product developers and designers for future product improvements. This research paper discusses the user journey mapping approach for analyzing product reviews. It proposes a method for structuring large amounts of user reviews and putting them on the journey map, classifying touchpoints, pain points, and product advantages. Machine learning algorithms on Apple Earpods Max noise-canceling headphone reviews are used to classify user-generated product reviews and validate the journey map. Created journey map showed a positive potential for the given approach to make sense of large amounts of user-generated content and give quantifiable proof of a user journey map.

**Keywords:** User experience knowledge · User journey map · User review analysis

## 1 Introduction

The user journey map is a creative method for a quick entry into complex UX projects. In a short time, it allows learning about relevant user processes and identifying and planning necessary UX activities, even before entering the user research phase [1]. The critical components of the user journey map are a business or product-specific touchpoints on X-axes and user thoughts, actions, and emotional states during them - labeled as valuable managerial insights for improvement on Y-axes [2].

A journey map often is created collaboratively with the group of stakeholders and subject matter experts in the interactive process led by the moderator [1]. This process is to identify areas of opportunities – often user pain points, which can later be solved and thus improve overall user experience and business offering [3]. However, the disadvantage of a user journey map created with the help of stakeholders is that it is time-consuming, relies on judgments made by individual experts [4], and may exclude other parties that may also have valuable input. Moreover, it is hard to evaluate the accuracy of a journey map filled in such away.

Ē. Nazaruka et al. (Eds.): BIR 2022, LNBIP 462, pp. 205–218, 2022.
https://doi.org/10.1007/978-3-031-16947-2_14

Various alternative data sources can be used to evaluate user experience and create a validated user journey map. For example, an eye-tracking device tracked users' behavior in e-commerce, and gathered data is used to fill the user journey map. Such created user journey map gives designers to create more personalized shopping paths and improved shopping experiences [5]. In addition, user interviews and surveys are a popular method to define journey map touchpoints [2, 6] and fill the journey map with relevant data [7]. However, these works have not considered user-generated content in online product reviews as the data source for user journey maps.

This research proposes how a UX expert-defined user journey map touchpoints can be filled with information based less on individual stakeholder input, with biases and limited knowledge, but on accurate user-generated content. This way, it could be possible to combine the best of both worlds – touchpoints based on expert creative input and content of activities, thoughts, and emotions would be based on actual user inputs about the product.

The rest of the paper is organized as follows: Sect. 2 discusses related prior work and the concept of the user journey map in detail. Section 3 describes the proposed method. Section 4 presents practical results with proposed methods in use. Finally, Sect. 5 concludes the user journey map used with this proposed method to structure user-generated reviews of a single product.

## 2    Framework

### 2.1    Related Work

Product reviews are an essential factor for customers when making purchase decisions. And not just the customers; for example, businesses are also interested in what their customers say about their products [8]. However, gathering qualitative information from users may be a tiresome, costly activity that requires much time. Therefore, researchers are interested in making use of user-generated content. Particularly – product reviews on platforms like Amazon, Twitter, Yelp, and more, as there is a lot of voluntarily written user feedback. Providing a method for analyzing this vast resource could change how product research is done, especially when using the user journey maps. For example, instead of typical resource-consuming user interviews, surveys, and focus groups, a user journey map of user wants, needs, and product challenges could be extrapolated from existing product reviews.

Extracting meaning from the user-generated product reviews is nothing new. Many types of research are dedicated to analyzing how to improve the product for the user. The Kano model and Quality Function Deployment (QFD) are the two most popular methods used in various combinations with other methods. In the literature review, Kano vs. QFD comparison authors state that the Kano model is a valuable tool for understanding customer needs and analyzing the effect of meeting customer needs on customer satisfaction levels. The QFD approach translates customer requirements into technical specifications and applies to product and service design [9]. While these are already established and self-proven methods, researchers use other approaches, such as sentiment classification, topic modeling, text categorization using faucets, prefixed user experience dimensions, text summarization, and many more.

While searching for relevant research that would use a user journey map to extrapolate user experience factors from user-generated content, one that came close was Ken Chinzinsky approach to measuring UX investments over time from user evaluated information. They developed a web tool that allowed users to evaluate touchpoints with the sales team by rating satisfaction levels with them. The result was displayed as a sentiment heatmap compared to previous years to see if a particular investment in UX has paid off or not [10]. Sentiment evaluation is present in most user-generated review analyses [11–13] as that defines whether the user's feedback was positive, negative, or neutral. Researchers like to compete in achieving sentiment values' with the most accurate classification method [14]. Although sentiment evaluation is typically represented as the probability of positive or negative sentiment, Steffen Hedegaard and Jakob Simonsen provided an alternative version. They extracted user sentiment and created usability and user experience associated vocabulary from the most informative word stems. They used extensive usability and UX categories such as Learnability, Errors, Enjoyment, Usability etc., from the literature review, classified the sentences accordingly, and extrapolated informative word stems to create usability dimension vocabulary [12]. Although they used Amazon game reviews as their input data, their vocabulary ended with multiple gaming jargon. There is potential for qualitative sentiment analysis, but we argue that this approach then should be re-used for each niche product separately as each product or product niche has its specific jargon and language.

Research shows that sentiment analysis by itself is not enough. It is worth combining it with other methods to understand user reviews better. Such an approach is demonstrated by Feng Zhou, who used Amazon product reviews, prepared them for topic modeling, did a sentiment analysis, and categorized reviews using the Kano model [15]. The result was categorized customer needs based on satisfaction and dissatisfaction scores for future product improvements. Bai Yang covered another approach to achieving user experience-related knowledge for future improvements. They used facet method by categorizing Amazon product reviews in product, situational, cognitive and sentiment facets. With this approach, similarly to usability dimension vocabulary – researchers were able to uncover specific language segments used to describe features that increase or decrease user experience in a particular product or product category [16].

In this research, a user journey map approach is proposed to evaluate user experience factors in a linear user-product interaction manner, where key touchpoints are defined by the needs and interests of the business and the observer himself.

## 2.2  User Journey Map and Its Design

Journey mapping is a popular method used in service design, service blueprinting, customer experience, user experience, and other related fields to map out user activities to reach a goal. The primary purpose of a journey map is to collect and analyze data obtained through customer research and use it to visualize the experiences grouped by the user or customer types. Meaning that each unique experience is not evaluated individually, but the aim is to find commonalities among various customers to create a sub-set of similar experiences [17]. The goal of the journey map is not only to translate the lived experience of the customers visually but also to improve solution design and create new offerings [18].

Kokins in his literature overview on customer-centric journey mapping points out the importance of the experiential aspect in journey mapping. The experiential aspect is often ignored in service design or blueprinting [7] while it is essential in computer science and digital products. Experiential aspects show better user response and significant satisfaction with information systems, digital products, and service use [19]. This concludes that it is not enough to show touchpoint use cases and actions taken by the user in service blueprints. Journey maps propose more dimensions for improving product experiences, and to make future product improvements out of them, it requires some way of representing experiential factors between user and product. Therefore, it is not enough to show user actions and use cases in touchpoints. Where touchpoint refers to any time that a user meets an on or offline experience, service, or showcase related to business or product along their customer journey. Typical journey map design consists of X and Y axes. The Y-axis of the map should represent aspects of the journey that can be used to gain valuable managerial insights for improvements – and they should be inextricably linked to the touchpoints, as doing so connects the journey map process with service improvement and innovation possibilities [20]. Some might argue that touchpoints – the X-axis should be defined through user research on what is defined as necessary by the user [2]. As there is no evidence of the improvement of this approach, we argue that touchpoints depend on what is essential for the observer in the context of future use of the journey map. Journey maps are often represented as tables, as this layout proposes an easier way to glance at a large amount of content (see Fig. 1).

In this research, the UX expert observer defines user journey map touchpoints based on his subject knowledge. Of course, touchpoints can be changed, and algorithms can be re-trained to adapt to the changes, so this aspect does not impact the research method. The main proposal of this research is that it is the aim to structure product reviews in the user journey map method, fill the content of the journey map using reviews content and create a valid user journey map.

**Table 4.** The HPM customer journey map: Post-service

| | | Touchpoint | | |
| --- | --- | --- | --- | --- |
| | | Recommend a store to others | Talk to others about purchases | Plan on returning to HPM |
| Strategic Action | Mall shopper requirements | To have a positive attitude toward mall retailers. | To have a positive attitude toward recent purchases at mall. To be able to obtain memories during their shopping experiences. | Mall shoppers need reasons (products, services, information, entertainment) to return. |
| | Employee actions | Station customer ambassadors at help desks near entrance/ exist doors to encourage shopper feedback. Encourage mall tenants to display return policy. | Encourage mall shoppers to share purchase and activity information with others via social media. Provide shoppers with two screening areas that all them to post pictures. Give shoppers an incentive to use the mall's hashtag. | Mall marketing manager, social media manager, and operations plan email and mobile application strategies that allow shoppers to receive real-time updates on shopping center activities, show times, links to retailers, and their specials. |
| | Employee support | Mall's social media director is responsible for all post-service, online communication with mall shoppers. | Mall's marketing manager, social media director, and operations are responsible for maintaining the two screening areas in the mall. | Marketing manager, social media manager, and operations must be knowledgeable on internet technology. |

**Fig. 1.** Example of the journey map [2].

## 3  Method

This research proposes a method to structure and fill the content of the user journey map from user-generated product reviews on Amazon. The approach is split into four main stages: exploring the data set and touchpoint definition, cleaning the dataset, testing and evaluating the machine learning algorithm, applying the best algorithm, and categorizing the reviews to finalize the user journey map (see Fig. 3). The proposed approach explores the available data set, which can be practically any user-generated content based on an identical product and shares similarities with what users do and refer to in the reviews. The observer then analyzes the dataset to identify relevant touchpoints based on his expert evaluation. For example - what information topics are repeating and would be worthy of being brought out for summarized analysis later.

In the second stage – the dataset is cleaned and prepared for training. Typically, user reviews are long and contain a large amount of information containing many touchpoints. If the whole multi-sentence review were classified as one – many product use cases would be lost. Because of that – reviews are separated into individual sentences, and sentences individually will be classified into a touchpoint (see Fig. 2). Each sentence is cleaned from stop words, hyperlinks, foreign (non-English) words, emails, emoticons, punctuation, double space, tabular space, etc. Descriptive nouns and verbs are extrapolated from each sentence to extract the core meaning of the sentence and ease content analysis using keywords. This is automated task using Python NLTK library and simple script that goes through each sentence and performs data cleaning.

In this case, machine learning will classify the sentences, and for that purpose, a training data set is needed. Expert knowledge is needed as he assigns sentences to the touchpoints in Fig. 2. It is possible to see that one sentence may refer to multiple touchpoints. Expert assigns one sentence to no more than three touchpoints as three was observed maximal amount touchpoints sentences usually had.

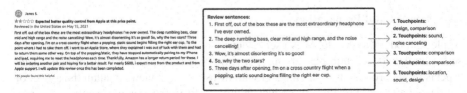

**Fig. 2.** Example of Amazon Earpods Max user review split into sentences – individual use cases and touchpoints.

In the third stage, algorithm training and evaluation take place to define which one will be used to classify the remaining dataset. Multiple machine algorithms are used to classify user review sentences to place them in user journey maps under matching touchpoints. It is hard to tell which algorithm would best suit this task. So multiple varying algorithms are proposed to be tested by taking a random sample from the training dataset and applying it and later comparing for the precision and recall, known as f1 score, among other algorithms.

The most successful algorithm is selected and applied to classify the remaining dataset in the fourth stage. Due to the complex nature of product reviews - one sentence can contain multiple references to more than one touchpoint - each sentence is classified into up to three touchpoints, and the probability values are represented for evaluation.

As noted in related works – sentiment analysis helps identify written text sentiment to identify further whether the sentence user wrote is positive, negative, or neutral. Also, user sentiment or feelings are essential evaluation factors in user journey mapping techniques. VADER (Valence Aware Dictionary and sEntiment Reasoner) is a lexicon and rule-based sentiment analysis tool specifically attuned to sentiments expressed in social media and is used to calculate the value of the sentiment.

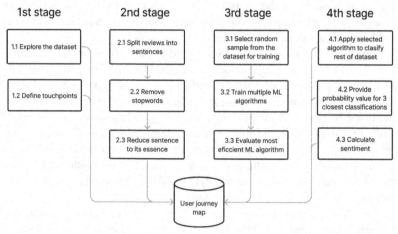

**Fig. 3.** Proposed four stage approach to create user journey map from available product reviews.

## 4  Data Analysis

### 4.1  Data Set and Preparation

Apple Earpods Max headphones user reviews on Amazon were chosen as input data set because there are multiple varying user reviews for this product as users have many aspects of the product they can interact with. Most of the reviews contain the aspect of the design and build quality to products ability to fit in the existing Apple ecosystem and, of course, the audio, noise-canceling aspect itself. In total, 48 reviews written in English were selected for the dataset containing. Only the user-generated text was used. No other information was used, like date of creation, rating 1–5, or whether other people found the review valuable.

UX expert performed an exploratory review of the available dataset to identify repeating topics, overall user concerns, and frequent use cases. That was done to define touchpoints that would be later used to classify review content. Selected touchpoints were – headphone connection, noise-canceling quality, battery charging, headphone case, comparison with other headphones, software-related issues, the location where the product is used, and opinion of the design and ecosystem of other Apple products headphones are used with. The training sample is created by manually assigning touchpoints with representative touchpoints by the UX expert.

In total, 588 review sentences are retrieved from the available starting set. Before algorithm training and classification can be started, each sentence is cleaned by removing stop words, hyperlinks, foreign (non-English) words, emails, emoticons, punctuation, double spaces, etc., and stored for training and classification. Additionally, for text-based user journey map – is an attempt to extract the essence of each sentence by stripping it down to nouns and verbs. Thus, showing the right keywords that are represented in later classified touchpoints.

Because a machine learning algorithm will be used, it is necessary to provide it with training and test data. The expert goes through the 588 reviews and assigns touchpoints to the sentences that match the representative touchpoints. For more straightforward data representation, classes to touchpoints are assigned as follows: connection – 0; noise-canceling – 1; battery charging – 2; case – 3; comparison – 4; software – 5; ecosystem - 6; location – 7; design – 8; audio – 9. The result of the second stage is a dataset prepared for training and testing machine learning and filling content in the user journey map (see Fig. 4).

| sentence | class | sentence_prepared | final_sentence |
|---|---|---|---|
| Every so often, I have to manually tell it to use these, but most of the time it switches smartly. | 0 | Every so often , I have to manually tell it to use these , but most of the time it switches smartly . | manually tell use time switches smartly |
| Finally, the AirPods sound better (by quite a bit, it's not even close) at 70 percent volume and above. | 9 | Finally , the AirPods sound better ( by quite a bit , it is not even close ) at NUMBER percent volume and above . | finally airpods sound better bit close number percent volume |
| While the sleeve does protect the metal earpieces of the headphones from getting scratched up, for the most part, it doesn't protect the headband and there is even a cut out on the bottom that leaves the headphones exposed to the elements and potential damage. | 3 | While the sleeve does protect the metal earpieces of the headphones from getting scratched up , for the most part , it does not protect the headband and there is even a cut out on the bottom that leaves the headphones exposed to the elements and potential damage . | sleeve protect metal earpieces headphones getting scratched protect headband cut leaves headphones exposed elements potential damage |
| No brick, for $550 you would think they would include a charging brick but all they give you is a Lighting cable that does NOT have a standard usb connection on one end. | 2 | No brick , for $ NUMBER you would think they would include a charging brick but all they give you is a Lighting cable that does NOT have a standard usb connection on one end . | brick number think include charging brick lighting cable standard usb connection end |
| I will first start off by stating that the noise cancelation, compared to the PXC 550-II is significantly better. | 1 | I will first start off by stating that the noise cancelation , compared to the PXC NUMBER II is significantly better . | start stating noise cancelation compared pxc number ii significantly better |
| It does not completely cancel out all noise however, it does a better job. | 1 | It does not completely cancel out all noise however , it does a better job . | completely cancel noise better job |
| I have not charged them yet and after 20 hours I still dont have to charge them, they are at 30% | 2 | I have not charged them yet and after NUMBER hours I still do not have to charge them , they are at NUMBER . | charged number hours charge number |
| The Digital Crown is exactly like what you will find on the Apple Watch, only bigger. | 8 | The Digital Crown is exactly like what you will find on the Apple Watch , only bigger . | digital crown exactly like find apple watch bigger |
| this is due to the metal design and the mass amount of microphones that they have located in various location on the headphones. | 1 | this is due to the metal design and the mass amount of microphones that they have located in various location on the headphones . | metal design mass microphones located location headphones |
| But pairing and switching between (Apple) devices is almost seamless and pretty near impossible to beat given their special support for their own headsets. | 0 | But pairing and switching between ( Apple ) devices is almost seamless and pretty near impossible to beat given their special support for their own headsets . | pairing switching apple devices seamless pretty near impossible beat given special support headsets |
| No doubt they are as good if not better sounding than my Bose. | 4 | No doubt they are as good if not better sounding than my Bose . | doubt good better sounding bose |
| When you use the transparency mode these microphones really shine! | 1 | When you use the transparency mode these microphones really shine ! . | use transparency mode microphones shine |
| If you own a couple of Apple devices, say a Mac and an iPhone, the Max's convenience makes them even more compelling. | 6 | If you own a couple of Apple devices , say a Mac and an iPhone , the Max s convenience makes them even more compelling . | couple apple devices mac iphone max convenience makes compelling |
| I don't watch movies on my iPhone or even iPad. | 6 | I do not watch movies on my iPhone or even iPad . | watch movies iphone ipad |
| Very comfortable and music quality is superb for Bluetooth/wireless since I listen to Spotify it is the best quality you will get out of wireless in my opinion. | 5 | Very comfortable and music quality is superb for Bluetooth/wireless since I listen to Spotify it is the best quality you will get out of wireless in my opinion . | comfortable music quality superb bluetooth wireless listen spotify best quality wireless opinion |
| The Bose sound bette at 60 percent volume and below. | 9 | The Bose sound bette at NUMBER percent volume and below . | bose sound bette number percent volume |
| You're better off buying beats headphones. | 4 | you are better off buying beats headphones . | better buying beats headphones |
| When this mode is active, and no music is playing, I can clearly hear myself as if I did not have the headphones on at all! | 1 | When this mode is active , and no music is playing , I can clearly hear myself as if I did not have the headphones on at all ! . | mode active music playing clearly hear headphones |

**Fig. 4.** End of the second stage – snapshot of dataset that is prepared for machine learning.

## 4.2  Training and Algorithm Selection

In this case no machine learning algorithm is selected by default for the text classification task. Instead, multiple varying algorithms are trained and tested to select most accurate one, that further will be used to classify full dataset. A XGBC Classifier, SVC, Logistic Regression, SGDC Classifier, Multinomial NB, Gradient Boost classifier, Random Forest classifier, Ada Boost classifier, Extra trees Classifier, MLP classifier and KNieghbour classifier are used to define which of them performs the best and will be used further.

20% - 117 randomly selected sentences from the data set are selected with given classification classes used to train each of the algorithms. Then, another random 117 sentences from the dataset are used to test the classification algorithm. Finally, the algorithm classification test result is compiled in Table 1.

**Table 1.**  Result of algorithm classification test.

| Name | F1 score | Precision | Accuracy | Recall |
|------|----------|-----------|----------|--------|
| XGBClassifier | 0.5467 | 0.5786 | 0.5593 | 0.5593 |
| SVC | 0.6308 | 0.6454 | 0.6440 | 0.6440 |
| LogisticRegression | 0.5990 | 0.5974 | 0.6101 | 0.6101 |
| SGDClassifier | 0.6269 | 0.6863 | 0.6355 | 0.6355 |
| MultinomialNB | 0.5857 | 0.5808 | 0.6016 | 0.6016 |
| GradientBoostingClassifier | 0.1917 | 0.1581 | 0.2966 | 0.2966 |
| RandomForestClassifier | 0.6191 | 0.6541 | 0.6440 | 0.6440 |
| AdaBoostClassifier | 0.5117 | 0.5283 | 0.5254 | 0.5254 |
| ExtraTreesClassifier | 0.5899 | 0.6192 | 0.6016 | 0.6016 |
| MLPClassifier | 0.4841 | 0.4755 | 0.5084 | 0.5084 |
| KNeighborsClassifier | 0.55722 | 0.5774 | 0.5762 | 0.5762 |

Testing results established that SVC algorithm accuracy was the highest and f1 score – 63,08%. In classification matrix it's possible to see which touchpoints were classified with the highest accuracy - noise cancelling, audio, the design had the largest training set examples opposing the one where accuracy was low – software, connection, case (see Fig. 5). Theoretically, classification accuracy could be higher if the training set were balanced, and each touchpoint would have an equal number of classifications. Unfortunately, an imbalance in the training set occurred due to the training set was classified manually.

## 4.3  User Journey Map Finalization

Based on dataset observation – one sentence can contain multiple references to multiple use cases and belong to multiple touchpoints in the user journey map. To not miss out on potential knowledge - each sentence is classified into up to three touchpoints, and its

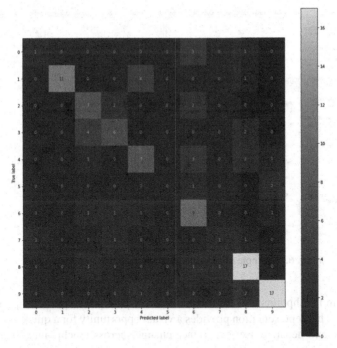

**Fig. 5.** End of the second stage – dataset is prepared for machine learning.

probability value is provided for observers' evaluation. Furthermore, a sentiment classification is initiated to know whether a sentence in a touchpoint was mentioned with a positive, neutral, or negative sentiment. To achieve that - VADER (Valence Aware Dictionary and sEntiment Reasoner, is used to assign each sentence with sentiment evaluation. An extracted essence of each review sentence is added to the journey map alongside the sentiment evaluation. In total, 588 sentences previously prepared for classification as described above are classified using the SVC algorithm. Each sentence is stripped to its keywords of nouns and verbs, classified up to three representative touchpoints, and given a sentiment evaluation using the VADER method. The resulting classification information with classified touchpoints, sentiments of the sentence and classification probability of three touchpoints is visualized in Fig. 6 based on user-generated reviews on Amazon and Apple Earpods Max Noise cancelling headphones.

### 4.4 Analysis

User journey maps typically, but not necessarily, are a visual artifact that visually represents users' experience with the product in the given touchpoints. The X-axes of the journey map represent previously selected touchpoints, and on Y-axes - represent aspects of the journey that can be used to gain valuable managerial insights for improvements that should be inextricably linked to the touchpoints. This visual representation (see Fig. 7) uses quantitative values from classified sentences to display average positive and negative affection per touchpoint. In addition, quantitative values of the 4 most positive

| Final sentence | Sentiment | Value | Ecosystem | Location | Design | Audio | Noise canceling | Case | Charging | Comparison | Connection | Software |
|---|---|---|---|---|---|---|---|---|---|---|---|---|
| connecting instantly iphone picked having use typical bluetooth prompt accept connection worth enslaving aluminum ears | positive | 6.88 | x 0.14 | | | | | | | x 0.14 | x 0.29 | |
| thought maybe apple worked kind voodoo music files sound better limited standard bluetooth connection | negative | 8.34 | x 0.08 | | | x 0.07 | | | | | x 0.57 | |
| active noise canceling anc airpods max anc sony headphones best | positive | 25.7 | | | x 0.02 | | x 0.86 | | | x 0.05 | | |
| lots people big deal power button talk draining battery case | negative | 8.1 | | | x 0.06 | | | x 0.10 | x 0.64 | | | |
| need ensure devices updated latest operating system order fully enjoy features | positive | 19 | x 0.19 | | | x 0.09 | | | | | | x 0.20 |
| seamlessly switch macbook iphone ipad apple tv watched apple tv dinner sound incredible | positive | 22.5 | x 0.36 | | | x 0.03 | | | | x 0.45 | | |
| travel lot charge | neutral | 100 | | x 0.30 | | | | | x 0.18 | x 0.18 | | |
| real advantage found work place | positive | 12.5 | | x 0.63 | x 0.06 | | | | x 0.05 | | | |
| need pick party case traveling afraid getting smushed backpack | positive | 9.4 | x 0.07 | | | | | x 0.51 | | x 0.08 | | |
| likely aluminum ear ups acting heat dissipation device airpods max cool condense water vapors getting drawn warmer air near ear inside ear cups porous mesh memory foam material ear pads | positive | 8 | | | x 0.59 | x 0.05 | | | x 0.07 | | | |

**Fig. 6.** Example of classified sentences for user journey map.

advantages in touchpoints and the 4 most negative pain points per touchpoint are also represented. This representation provides a visual opportunity for a quick assessment of how positive and negative user experience changes across touchpoints, displayed as a line chart, and assesses the most impactful advantages and pain points in the touchpoint.

Touchpoints are selected and organized to represent the linear product usage cycle as close as possible. First, the user opens the headphone box, has first contact with the product's design, establishes a connection with the device, and plays audio. The next observed thing is the noise-canceling and overall comparison with previously owned similar headphones. Then, use the product in various locations, traveling and putting headphones in their case, charging them, connecting with the rest of the Apple ecosystem products, and seeing where the software gets mixed in.

The resulting approach provides a user journey map with quantitative information and validation of the user's review belonging to one or three touchpoints and positive or negative sentiment. This approach differs from typical user journey map development, where the facilitator and project stakeholders create user journey maps based on stakeholder expertise. Such an approach was criticized by Rosenbaum, claiming that it lacks objectiveness [2]. While this research aim was not to improve the objectiveness of the user journey map but propose a way that the user journey map can be used to structure and make sense of user reviews – the qualitative approach of a sentence belonging to a touchpoint does precisely that. Furthermore, by classifying sentences into touchpoints, we gain probability scores to evaluate the probable accuracy of users' feedback.

While proposed approach is valuable and reduces manual work, the precision of classification – 64% should be higher and current implementation would provide noticeable amount of wrongly classified touchpoints. Furthermore, despite of attempt to automate creation of user journey map and classification of touchpoints, there are still manual work that needs to be done. Such as definition of touchpoint and decision of which

| | Design | Connection | Audio | Noise cancelling | Comparison | Location | Case | Charging | Ecosystem | Software |
|---|---|---|---|---|---|---|---|---|---|---|
| Emotionally affected (40, 30, 20, 10) / Feeling neutral (0) | | | | | | | | | | |

— Positive affection   -- Negative affection

| | Design | Connection | Audio | Noise cancelling | Comparison | Location | Case | Charging | Ecosystem | Software |
|---|---|---|---|---|---|---|---|---|---|---|
| | amazing headphones | easy setup excellent connection apple devices | amazing sound quality | excellent anc | overall money value good sound quality easy buy | trying w/in beauty playyard prefer look zoom calls | smart case helps maintain battery life use | great battery life | easy integration apple eco system | iphone sure updated latest ios help phone truly recognize headphones update firmware |
| | overall excellent headband head comfortable | connects seamlessly ipad iphone almost good | headphones sound great | noise cancelling awesome | headphones amazing worth penny you lifetime | love little world working gym listening noises machines gym people lot | nice material provide protection | battery holds strong | connection phone ipad perfect | wish accompanying app enabled intuitive eq control settings like degrees anc better monitoring battery life standby |
| **Advantages** | controls perfect | pairing switching apple devices seamless pretty near impossible beat given special support headsets | great sound quality functionality | transparency mode awesome | better buying beats headphones | drive yessir living wear number hours kidding far beat | sleeve protect metal earpieces headphones getting scratched protect headband out leaves headphones exposed elements potential damage | battery life far good | better perfect ios devices going forth phone ipad seamless pretty amazing better | honestly heavily happily invested apple ecosystem write type over priced macbook probably skipped pad buy benefits chips sweet macos too only features invasion deal |
| | pretty good controls connectivity | like airpods pro bluetooth better tried | sound absolutely wonderful | passive mode great work talking phone | better sounding box sure | real advantage found work place | small qualm case material unbox max strong unpleasant polymer type odor goes away quickly | battery life good buttons right cup feel work great | max avoids greatly pairs apple tv watching movies | need ensure devices updated latest operating system order fully enjoy features |
| **Pain points** | heavy real comfortable | normal bluetooth pairing utility annoying depending devices frustrating | sound okay bass lacking volume shockingly low overall disappointment | difficult finding different anc mode suspect | works better w/in bose number gets lost easy reset | times simply infuriating minutes writing trying answer kitchen phones desk room went airpods | case terrible | big miss power | overall disappointed instant switching involving macbook pro | got latest firmware update |
| | slight design decisions annoying | key paired bluetooth devices miss features | fault bass tosses little nuance | unfortunately loud distraction work call sticky noise cancel helps focus | hopping better am constantly got stuck wrong device way running bluetooth send laptop | trying w/in beauty playyard prefer look zoom calls | case embarrassment | battery life horrible | connecting ios devices excellent macbook pro great | probably fixable future software update |
| | said warmth bother feel uncomfortably hot | device switching apple devices flawless thought headphones hung device | number percent volume boss day begins lower bass response avoid distortion | apple avoid effort sort synthetic silence hiss noise cancelling | boss warm apple | | case everybody hates | charges lightning cable | better perfect ios devices going forth phone ipad seamless amazing better | things apple change software update imagine addressed future iterations |
| | reviews complaining weight | thought maybe apple worked bland voodoo music files sound better limited standard bluetooth connection | ok way sound good | But when not playing audio, the ANC causes too much pressure on my eardrums | doubt good better sounding boss | | case w/f | hate apple stopped including ios wall adapters recent products | connection macbook pro phone seamless gets lost | |

**Fig. 7.** Final user journey map.

touchpoint to use when classified probability is equal requires expert input. In a future works this would require improvements.

Despite the shortcomings, where typical user journey map creation is based on loads of manual work, discussions, and manual user data research - the proposed approach provides benefits of automation. An abundance of online product reviews can be used instead of doing dedicated user research before developing a user journey map. Gathering data from online sources and setting up the training dataset can take some time, but as a result, it takes much less time to classify, summarize, and place the correct text under the right touchpoint. Creating a user journey map manually poses the same challenges (gathering data, deciding on touchpoints, establishing sentiment, etc.). In addition, it can't be scaled to 10,000 user-generated sentence reviews, unlike the proposed solution, for whom that is a matter of seconds.

It is essential to point out that the precision of classification – 64% could be higher, as shown in text classification algorithm research, where accuracy can reach up to 89% [14]. In addition, future work could be improved by an equally large touchpoint training sample size.

## 5   Conclusion

The proposed approach proved capable of processing, cleaning, and classifying user reviews into sentences, and touchpoints, establishing sentiment, and providing a user journey overview. This approach allows to classify large amounts of user-generated product reviews using the user journey map method that has not been tried before. This method would allow the expert to avoid the manual task of going through user reviews and assign them to respecting touchpoints in user journey map. That could be achieved with one machine learning algorithm. Due to the nature of classification algorithms, the proposed approach provides probabilistic quantitative values of content in the user journey map. Comparing that to normal user journey maps is considered a new addition and can be used to validate the user journey map itself. The benefit of the given approach is that it is easily scaled, can be adjusted to other products, not just Apple Earpods Max, and holds only computational limitation to processing many user reviews.

There are no similar user journey map classification attempts made from Amazon reviews. Therefore, it is hard to compare results. However, it is possible to compare the advantages of this approach. For example, the observer can define touchpoints, and instead of using stakeholder or user interviews to fill out the user journey map manually – this can be done faster using a more considerable amount of voluntarily provided information by real users. In addition, touchpoints of the journey map provide simple.

categorizing, filtering, and overviews of user-product interaction, and sentiment evaluation provides polarity of user's emotions. Of course, the classification accuracy could be higher and provide more precise results, but that can be improved with better training data set.

This approach can also be used to mine product relationships within the product ecosystem. In this journey map, one of the touchpoints was – an ecosystem that represented Earpods Max connected use with other products from the Apple ecosystem such as iPhone, Mac, Apple TV, etc. This option, with further research, can be used to evaluate reviewed products in the larger context of product ecosystems in which the product is used.

# References

1. Endmann, A., Keßner, D.: User journey mapping – a method in user experience design. i-com **15,** 105–110 (2016)
2. Rosenbaum, M.S., Otalora, M.L., Ramírez, G.C.: How to create a realistic customer journey map. Bus. Horiz. **60,** 143–150 (2017)
3. Xiong, Z.Y., Xiao, J.X.: Research of smart door lock design based on an extended Kano model and user journey map. In: Proceedings - 2021 2nd International Conference on Intelligent Design, ICID 2021. Institute of Electrical and Electronics Engineers Inc., pp. 255–259 (2021)
4. Väinämö, O.: Refining the user journey co-creation processes, Master's Thesis, Laurea University of Applied Sciences (2016)
5. Tupikovskaja-Omovie, Z., Tyler, D.: Eye tracking technology to audit google analytics: analysing digital consumer shopping journey in fashion m-retail. Int. J. Inf. Manag. **59** (2021)
6. Alkhalisi, A.F.: Creating a qualitative typology of electric vehicle driving: EV journey-making mapped in a chronological framework. Transport. Res. F: Traffic Psychol. Behav. **69,** 159–186 (2020)
7. Kokins, G., Straujuma, A., Lapiņa, I.: The role of consumer and customer journeys in customer experience driven and open innovation. J. Open Innov. Technol. Mark. Complex. **7**(3), 185 (2021)
8. Vollero, A., Sardanelli, D., Siano, A.: Exploring the role of the Amazon effect on customer expectations: an analysis of user-generated content in consumer electronics retailing. J. Consum. Behav. (2021)
9. Ishak, A., Ginting, R., Suwandira, B., Fauzi Malik, A.: Integration of Kano model and quality function deployment (QFD) to improve product quality: a literature review. In: IOP Conference Series: Materials Science and Engineering. IOP Publishing Ltd., Bristol (2020)
10. Chizinsky, K.: Quantifying user experience in a sales journey (2018)
11. Yunus, N.H.M., Sampe, J., Yunas, J., Pawi, A., Jalil, M.I.A.: Performance comparison of micromachined antennas optimized at 5 GHZ for RF energy harvester. Indones J. Electr. Eng. Comput. Sci. **15**(1), 258–265 (2019)
12. Baudisch, P., Beaudouin-Lafon, M., Mackay, W.E.: Extended Abstracts of the 31st Annual CHI Conference on Human Factors in Computing Systems (2013)
13. Sindhu, C., Mukherjee, D., Sonakshi: A Joint sentiment-topic model for product review analysis of electronic goods. In: Proceedings - 5th International Conference on Computing Methodologies and Communication, ICCMC 2021. Institute of Electrical and Electronics Engineers Inc., pp. 574–578 (2021)
14. Bhavani, A., Santhosh Kumar, B.: A Review of state art of text classification algorithms. In: Proceedings - 5th International Conference on Computing Methodologies and Communication, ICCMC 2021. Institute of Electrical and Electronics Engineers Inc., pp. 1484–1490 (2021)
15. Ayoub, J., Zhou, F., Xu, Q., Yang, J.: Analyzing customer needs of product ecosystems using online product reviews. In: Proceedings of the ASME Design Engineering Technical Conference. American Society of Mechanical Engineers (ASME) (2019)
16. Yang, B., Liu, Y., Liang, Y., Tang, M.: Exploiting user experience from online customer reviews for product design. Int. J. Inf. Manage. **46,** 173–186 (2019)

17. De Keyser, A.: Understanding and managing the customer experience. Doctoral dissertation, Ghent University (2015)
18. Epp, A.M., Price, L.L.: Designing solutions around customer network identity goals. J. Mark. **75**(2), 36–54 (2011)
19. Badran, O., Al-Haddad, S.: The impact of software user experience on customer satisfaction. J. Manag. Inf. Decis. Sci. **21**(1), 1–20 (2018)
20. Ņikitina, T., et al.: Competences for strengthening entrepreneurial capabilities in Europe. J. Open Innov. Technol. Mark. Complex. **6**(3), 62 (2020)

# Author Index

Printed in the United States
by Baker & Taylor Publisher Services